Cancer: The Outlaw Cell

Cancer: The Outlaw Cell
Second Edition

Richard E. LaFond, EDITOR

American Chemical Society, Washington, DC 1988

Library of Congress Cataloging-in-Publication Data

Cancer: the outlaw cell/Richard E. LaFond, editor.—2nd ed.
p. cm.

Includes bibliographies and index.

ISBN 0-8412-1419-0. ISBN 0-8412-1420-4 (pbk.)

1. Cancer—Research. 2. Cancer cells. I. LaFond, Richard E.

[DNLM: 1. Neoplasms. QZ 200 C2183]
RC267.C37 1988
616.99'4—dc19
DNLM/DLC
for Library of Congress 88-14517
CIP

PRINTED IN THE UNITED STATES OF AMERICA

Contents

INDEXES

About the Editor

Richard E. LaFond of Monson, Massachusetts, had the original inspiration to publish a series of articles on cancer research and its applications that would appeal to the specialist and layperson alike. He selected the authors, suggested graphic and editing approaches, and helped to promote the series when it became a reality.

These articles were gathered together in 1978 for the first edition of *Cancer: The Outlaw Cell*. Because of the popularity of that edition and particularly because of the striking advances in cancer research during the past ten years, Dr. LaFond decided to organize a second edition. Some of the chapters reflect ten years of progress from the point of view of the original authors. Other chapters delve into new topics that have become vital in the field of cancer research as it exists today.

Dr. LaFond is a member of the Research Faculty at the University of Massachusetts–Amherst, where he is conducting experiments in cell biology. Earlier, he was a research fellow in the Department of Biological Chemistry and Molecular Pharmacology at the Harvard Medical School. He also worked in the Department of Chemistry at Amherst College and the Clinical Laboratory at Providence Hospital in Holyoke, Massachusetts.

Scientific publishing is part of his life. Dr. LaFond served on the Advisory Board of *Chemistry* for four years, contributed three articles to the Amateur Scientist Department of *Scientific American*, and wrote for numerous newspapers, periodicals, and professional journals.

He is a member of the Board of Corporators of the Monson Savings Bank and was one of the first members of the Monson Conservation Commission. In addition, he is a member of the Expansion Committee of the Monson Free Library and Reading Room Association.

Contributors

JOAN RILEY BURCHENAL. Memorial Sloan-Kettering Cancer Center, New York, NY 10021. A former Research Assistant at Sloan-Kettering, Mrs. Burchenal is currently a science teacher in Darien, CT. She was the recipient of the Presidential Award for Excellence in Science Teaching in 1985.

JOSEPH H. BURCHENAL. Memorial Sloan-Kettering Cancer Center, New York, NY 10021. Dr. Burchenal has been at Sloan-Kettering as Attending Physician and Director of Clinical Investigation since 1946. Now an Emeritus Member of the Center and of the Cornell University Medical College, he has been a special consultant of the National Cancer Institute and Chairman of the Clinical Trials Committee of the Division of Cancer Treatment.

VINCENT T. DeVITA, JR. National Cancer Institute, National Institutes of Health, Bethesda, MD 20892. Dr. DeVita has been Director of the National Cancer Institute since his Presidential appointment in 1980. He is cited frequently for his accomplishments in the development of curative chemotherapy of several adult lymphomas, and for his role in the development of similar treatments for ovary and breast cancers.

JOHN L. FAHEY. Department of Microbiology and Immunology, The Center for Interdisciplinary Research on Immunology and Disease, University of California, Los Angeles, CA 90024. Dr. Fahey is Director of the Center for Interdisciplinary Research on Immunology and Disease, and Professor of Immunology and Medicine, UCLA School of Medicine. He was formerly Chief of the Immunology Branch, National Cancer Institute, Bethesda, MD.

ISAIAH J. FIDLER. Department of Cell Biology, University of Texas System Cancer Center, Houston, TX 77030. Dr. Fidler, Department Chairman and Director of the Division of Interferon Research, has contributed pioneering research studies into the properties of metastatic cancer cells. He has served as head of the Biology of Metastasis Section, Frederick Cancer Research Center, and has been President of the American Association for Cancer Research.

JUDAH FOLKMAN. Department of Surgery, Harvard Medical School, Boston, MA 02115. Dr. Folkman, Professor of Anatomy and Cellular Biology at Harvard Medical School, was Surgeon-in-Chief at Boston Children's Hospital for 14 years. He has been honored for outstanding clinical teaching and for his pioneering research into angiogenesis.

ELAINE V. FUCHS. Department of Molecular Genetics and Cell Biology, University of Chicago, Chicago, IL 60637. Dr. Fuchs, an Associate Professor at the University of Chicago, has a long record of research in cell biology. She received the Presidential Young Investigator Award and has been included in *Science Digest*'s list of the nation's 100 brightest scientists under the age of 40.

BRIAN C.-S. LIU. Department of Pathology, Henry Ford Hospital, Detroit, MI 48202. Dr. Liu, Director of Tumor Cell Biology at Henry Ford Hospital, did postdoctoral research in tumor immunology at the UCLA School of Medicine. His research interests are the role of oncogene activation and expression, and the mechanisms of such activation and expression in tumor progression.

ARIEL C. HOLLINSHEAD. Division of Hematology and Oncology, George Washington University Medical Center, Washington, DC 20037. A Research Oncologist and Professor of Medicine, Dr. Hollinshead was the first to isolate, purify, and identify tumor-associated antigens. She is codiscoverer of TAA specific active immunotherapy for cancer patients. Her work in immunotherapy has recently led to dramatic improvements in therapy for advanced lung cancer patients.

FREDERICK MEINS, JR. Friedrich Miescher-Institut, Basel, Switzerland. Dr. Meins, a New York native, has been a Scientific Group Leader at the Friedrich Miescher-Institut since 1980. He has taught at Princeton University and the University of Illinois at Urbana. His major research interests are regulation of tumor development and cell commitment in plants.

MARGARET L. KRIPKE. Department of Immunology, University of Texas System Cancer Center, Houston, TX 77030. Dr. Kripke, Chairman of the Department of Immunology, is internationally recognized for her work on the immunological effects of ultraviolet radiation. She is working to establish a high-level basic research effort in cancer immunology among members of her department.

DIANA FURST NELSON. Radiation Oncology Center, Highland Hospital, Rochester, NY 14620. Dr. Nelson has taught Radiation Therapy at the University of Pennsylvania and the University of Rochester. She served as the Chairman of the Brain Subcommittee of the Radiation Therapy Oncology Group at the University of Pennsylvania and has held many advisory positions in the field of radiation therapy.

ARNOLD J. LEVINE. Department of Molecular Biology, Princeton University, Princeton, NJ 08544. Dr. Levine is Chairman and Professor of Molecular Biology at Princeton University. In addition to his teaching and research in the field of microbiology, he has served on many committees and professional groups dedicated to the advancement of cancer research.

ARTHUR B. PARDEE. Dana-Farber Cancer Institute, Harvard Medical School, Boston, MA 02115. Dr. Pardee is a Professor of Biological Chemistry and Molecular Pharmacology at Harvard Medical School. He is also Chief of the Division of Cell Growth and Regulation at the Dana-Farber Cancer Institute and Past President of the American Association for Cancer Research.

HENRY C. PITOT. McArdle Laboratory for Cancer Research, University of Wisconsin–Madison, Madison, WI 53706. Dr. Pitot is Director of the McArdle Laboratory, a Professor of Oncology and Pathology at the University of Wisconsin, Medical Director at Large of the National American Cancer Society, and Past Chairman of the National Cancer Advisory Board. His personal research is on the biological pathology of carcinogenesis and enzymatic control mechanisms.

RUTH SAGER. Dana-Farber Cancer Institute, Harvard Medical School, Boston, MA 02115. Dr. Sager is Chief of the Division of Cancer Genetics at Dana-Farber Cancer Institute and Professor of Cellular Genetics at Harvard Medical School. Before starting the Laboratory of Cancer Genetics, she spent a year in London studying mammalian cell culture and cell genetics as a Guggenheim Fellow at the Imperial Cancer Research Fund Laboratory.

WILLIAM D. POWLIS. Department of Radiation Therapy, School of Medicine, University of Pennsylvania, Philadelphia, PA 19104–6934. Dr. Powlis is an Assistant Professor of Radiation Therapy at the Hospital of the University of Pennsylvania. He treats patients with hematologic malignancies and brain tumors. His research topics include clinical mathematical modeling of radiation treatment planning and megavoltage image processing.

ROBERT L. ULLRICH. Biology Division, Oak Ridge National Laboratory, Oak Ridge, TN 37831. Dr. Ullrich is a Senior Scientist and Head of the Radiation Carcinogenesis Unit, Biology Division, at the Oak Ridge National Laboratory. He has been a part of the Oak Ridge research effort since 1974 and an Associate Editor of *Radiation Research* since 1983. Professional honors include a Research Award from the Radiation Research Society.

JAMES G. RHEINWALD. Harvard Medical School, Boston, MA 02115. Dr. Rheinwald is an Associate Professor of Physiology and Biophysics at the Harvard Medical School. He has been granted a Faculty Research Award by the American Cancer Society. His main research interests are mechanisms of carcinogenesis and mechanisms regulating growth and expression of differentiated functions in epithelial cells.

ELIZABETH K. WEISBURGER. National Cancer Institute, National Institutes of Health, Bethesda, MD 20892. Dr. Weisburger, Assistant Director for Chemical Carcinogenesis, Division of Cancer Etiology, has been with the National Cancer Institute since 1949. Her main interests are chemical carcinogenesis and toxicology. She has been a consultant for various government agencies and has served on various editorial boards.

PHILIP RUBIN. Department of Radiation Oncology, University of Rochester Cancer Center, Rochester, NY 14642. Dr. Rubin is Chairman of the Department of Radiation Oncology at the University of Rochester Cancer Center. He has received awards for his achievements from the American Society for Therapeutic Radiology and Oncology and the Royal College of Radiology (London).

Preface

The first edition of *Cancer: The Outlaw Cell* originated in a series of articles first published in 1977 by *CHEMISTRY* magazine, a former American Chemical Society publication that was directed toward beginning science students. At the time I was a member of *CHEMISTRY*'s Editorial Advisory Board. My suggestion that the magazine develop a series of articles on cancer was enthusiastically received by the magazine's staff and especially by Theodor Benfey, Editor-in-Chief. Dr. Benfey appointed me Guest Editor for the series.

The articles on cancer in *CHEMISTRY* and the resulting book were awarded several commendations. The first edition, later published in Arabic and Italian, has been among the most popular items on the ACS booklist for many years.

This new edition continues the tradition set by the first. It consists of articles that have been contributed by leading scientists on some of the most promising aspects of basic cancer research and clinical treatment. Each article has been written in a graphical style that is easily understood by students and the interested public. The underlying theme stresses fundamental principles of biology, how these concepts currently influence experimental study of cancer cells in the laboratory test tube, and how this experimentation influences the treatment of patients.

I am particularly grateful for the early support of Dr. Benfey and the encouragement I received from two present authors, Arthur B. Pardee and

Arnold J. Levine, who were my teachers at Princeton University. The popularity of the first edition was the impetus for the second edition, and the book has received the expert attention of Colleen Stamm, assistant editor, and Barbara Libengood, production specialist. I am deeply indebted to the authors and the ACS Books Department for producing a book that will surely inform and challenge all its readers.

RICHARD E. LAFOND, Editor
Monson, Massachusetts
March 20, 1988

Foreword

Vincent T. DeVita, Jr.

Major advances have occurred recently in the understanding, prevention, and treatment of cancer. These changes are apparent when comparing the first edition of this book, published in 1978, with today's edition.

In 1971, when the National Cancer Act was passed, most realistic scientists would have said that they hoped that more basic research would make it possible to accelerate progress toward an understanding at the molecular level of what makes a normal cell malignant. They would have said that if there were to be hope of preventing the disease—a first priority then and now—precise information at the molecular level would be necessary. The alternative would be to exhaust resources in random attempts to control cancer by making alterations in the environment.

Sixteen years ago, the cancer cell was still a black box. Support for basic research made it possible to lift the lid off this black box and expose the workings of the cell for all to see. We now can conceptualize what is going on in the cancer cell. We see it not as a random invading enemy, more dangerous than bacteria and viruses because of a capacity to circumvent any attempt at control, but as an aberration of a normal process–that is, the process of growth, development, and differentiation that follows upon the fertilization of the egg.

The outlaw cancer cell, in its disordered frenzy, appears to be trying to recapitulate phylogeny. It tries to make a whole human being, if you will, by using the normal genetic programs involved in growth and development, expressing genes, many of which we now know as oncogenes. These processes, extremely dangerous to a fully formed adult, have appropriately been shut down in most of us. When they are switched back on by a variety of interactions with the environment, we develop cancer.

In its undifferentiated form, the cancer cell effectively retains the capacity to express the most dangerous of these programs, the capacity to grow in situ and to travel and grow in secondary sites. We know this process as metastasis; in the end, it is lethal for most cancer patients. The metastatic process also appears to be under genetic control and to be an aberrant form of the normal cell migration that occurs during embryogenesis.

Trying to interfere with a random process is often like looking for a needle in a haystack. With our new knowledge of the molecular biology of the cell, however, we can begin to devise means to intercept, interfere with, or reverse a predetermined (albeit extraordinarily complex) series of programs that are the essence of life itself. This challenging but plausible exercise may well lead to the control of cancer, as well as yield staggering information on developmental biology that will be useful to all of medicine.

How the deeper appreciation of the malignant process came about is depicted in several chapters of this book. This revelation has happened so quickly that its magnitude is often unappreciated. Recently, when speaking to an international group of cancer researchers, I pictured myself somehow returning to 1971 and speaking to a similar group of scientists. In a sense, I would open the book, flip ahead, and "tell them how the story has come out so far." They would find it fascinating. I would tell them that, as a result of our investment in basic research, we would

- know about reverse transcription;

- be able to cut DNA into workable fragments and sequence it as a matter of routine;
- insert precise genetic constructs into bacteria and express specific genes;
- make gene products in quantity and study them;
- build an industry to produce and sell gene products, and use them in the diagnosis and treatment of cancer;
- discover that RNA is spliced after transcription and assists in its own splicing;
- discover that gene rearrangement regularly occurs; and
- discover that monoclonal antibodies could be made by fusing normal lymphocytes and malignant cells.

Most of all, I would tell them that these tools would uncover the presence and expression of a cascade of dominant genes, called oncogenes, in cancer cells. Oncogenes play an important, highly conserved role in the control of growth and differentiation alluded to earlier. Their misbehavior apparently leads to the aberrant growth we know as cancer.

In addition, I would tell them that there are also recessive genes whose role in this cascade may even be more important, and that this will be an important step in identifying persons at special risk of getting cancer.

I would tell such an audience of plans to alter germ cells to construct transgenic mice to test our ability to prevent the expression of these critical genes. I might explain that such transgenic mice are the most precise in vivo model ever developed to identify the means to prevent cancer.

If I could go back in time to speak about these things, people might smile in disbelief. Undoubtedly they would be both amazed and a little skeptical.

I might go further and describe a disease called AIDS. It would be a source of considerable concern to the scientists in 1971. But I believe they would say, "Certainly, invest in research on retroviruses." In addition to the valuable "side trips" into molecular biology, this knowledge would fill the desperate need to understand the life cycle of retroviruses, in order to think about how to control AIDS.

I think the scientists would also say, "Invest in basic studies of immunology, for we have only the barest inkling that there is more than one class of lymphocytes, let alone the tools to recognize and classify them. Nor any inkling of how they communicate with each other, recognize, and kill pathogens, but not our own cells. Nor the faintest idea of how we can generate such a diversity of responses to the millions of antigens to which we are exposed in our environment."

Fortunately, we did all of these things. This is why we can speculate now about the "intent" of the outlaw cell and why we can begin to think of a world without cancer. This is also why this book and others reflecting today's state of knowledge about cancer have become so interesting.

The writing of a foreword provides a valuable opportunity for reflection. Not only has research been productive, but the results of the research have advanced clinical practice. The cancer patient is infinitely better off today than in the past. Treatments are more effective. A major focus of the Cancer Act was to provide the resources to apply the results of basic research. In the mid-1970s a large number of clinical trials were established to test treatments, then new, of the majority of common lethal cancers. This effort has been extraordinarily effective, in that we have treatments that reduce mortality in the study populations with common cancers of the breast, colon, rectum, lungs, aerodigestive tract, sarcomas, and bladder. The challenge we are now facing is the rapid transfer of these new treatments into practice to reduce national mortality rates. This step is important to clinician and basic scientist alike, because the transfer machinery must be ready to take on the burgeoning, but pleasant, problem of integrating and applying the advances in molecular biology to the clinical care of patients.

Another point about clinical advances is worthy of emphasis. Quickly and almost unnoticed, treatment of most cancers has become far less morbid. Perhaps you have to have seen cancer treatment both two decades ago and now to appreciate this change. Equal or better survival results can be obtained today in breast cancer, colon cancer, rectal cancer, prostate cancer, head and neck cancer, bladder cancer, and bone and soft-tissue sarcomas, while preserving breasts, avoiding colostomies, avoiding impotence, avoiding disfiguring surgery, saving bladders, and sparing limbs. Over 385,000 patients will benefit in 1988 from these less morbid treatments, all of which have been developed in the last 17 years.

If we continue to discover new knowledge at the same pace, and learn to apply it efficiently, the next edition of *Cancer: The Outlaw Cell* will most certainly paint a vastly different picture of our ability to prevent and control major diseases like cancer and AIDS.

CHAPTER 1 Cancer—An Overview

Henry C. Pitot

Humans have been plagued by diseases throughout the history of civilization. In ancient times leprosy was the most dreaded disease; in Medieval and Renaissance Europe the scourge was the bubonic plague, or Black Death. A major killer often associated with extreme suffering in the 19th century was the White Death, or tuberculosis.

In the 20th century, using major advances in microbiology and pharmacology, medical science has conquered many of the infectious diseases that formerly destroyed large populations. With the discoveries of A. Fleming and S. Waksman, among others, and with marked improvements in medical care, our life expectancy has reached the biblical promise of three score and ten. But these benefits are not without a price: in this century cancer strikes fear in the hearts of many.

Through efforts of prominent scientists and several branches of the Federal Government, a panel of consultants was called together by the United States Senate in 1970 to report on a national program for the conquest of cancer. The financial backing requested by the panel, and ultimately signed into law, gave the greatest single impetus in our history to the search for the knowledge and understanding needed to control and eliminate cancer.

1420–4/88/0001$06.00/0

The panel report showed that cancer is in fact the primary health concern of the American people. In several surveys, approximately two-thirds of those questioned admitted fearing cancer more than any other disease.

More than 16% of all deaths in this country are caused by cancer—as a killer, it is second only to cardiovascular disease. Of 200 million Americans living in 1970, 50 million were destined to develop cancer and some 34 million would die of the disease. About half of all cancer deaths occur prior to age 65. In fact, cancer causes more deaths among children under age 15 than any other disease.

The panel pointed out that in 1969 the budget of this country, calculated on a per-person basis, allotted $410 for national defense; $125 for the war in Vietnam; $19 for the space program; $19 for foreign aid; but only 89 cents for cancer research. During the same year, cancer deaths equaled eight times the number of lives lost in all six years of the Vietnam war up to that time, and five and a half times the number of people killed in auto accidents. Further, the deaths exceeded the number of American servicemen killed in battle during all four years of World War II.

The report also indicated that the incidence of cancer is increasing, partly because the number of individuals in older age groups is increasing. However, the major factor in the growing cancer rate in the United States is the rising incidence of lung cancer, attributable almost entirely to tobacco smoking. The panel estimated that if Americans stopped smoking cigarettes, more than 15% of all cancer deaths in this country would be eliminated within two or three decades. The death rate from cancer would actually be decreasing, rather than increasing as it is now. The American Cancer Society estimated in 1985 that there would be 126,000 deaths from lung cancer in the United States during that year.

History of Cancer

All multicellular organisms can probably be afflicted by cancer, but only in this century has the general population become acutely aware of its significance. Paleopathologists have found cancerous lesions in dinosaur bones. Further, numerous reports of spontaneous and induced cancers in both plants and animals, such as planaria and fish, suggest that cancer has existed for most of the evolutionary period of life on Earth.

Egyptians were aware of cancer in human patients, and in one papyrus a glyph (Figure 1) clearly refers to a human tumor of the breast. In addition, autopsies of mummies have shown bone lesions and what appear to be other cancerous processes. Many types of tumors, such as stomach and uterine cancer, had been described by the fourth century B.C. Hippocrates coined the term, "carcinoma", which referred to tumors that spread and destroyed the patient. He called other growths "carcinos", including benign tumors, hemorrhoids, and other chronic ulcerations.

Figure 1. The symbol or glyph for "tumor" referring to a malignant tumor of the breast. This symbol was found in the hieroglyphics of the Edwin-Smith papyrus, dated to earlier than 1600 BC. In the papyrus, cancer of the breast, which could not be treated successfully, was distinguished from abscesses of the breast, also tumors, which could be treated successfully. The reader is referred to Breasted's translation of the document for further information (see Pitot, H.C., Fundamentals of Oncology, 3rd ed., Marcel Dekker, 1986, p. 3, for other details. The figure is reproduced from this text with permission of the publisher).

Almost 600 years later, Galen distinguished "tumors according to nature", such as enlargement of the breast with normal female maturation; "tumors exceeding nature", such as the bony proliferation occurring during the reuniting of a fracture; and "tumors contrary to nature", which today we call neoplastic growths. This distinction, proposed some 1800 years ago, is still reasonably correct. Galen also suggested the similarity between a crab and cancer.

Not until the 19th century, however, did physicians and scientists begin to study cancer systematically. Among the most prominent researchers of that era were Bichat, Muller, Pasteur, Laennec, Cohnheim, and Virchow. Many theories to explain the origin and development of cancer were proposed, but generally they can be classified as the irritation, embryonal, and infectious hypotheses.

The irritation hypothesis contained what little was known about chemicals and radiation as causes of cancer. Relationships of some chronic ulcerations to cancer appeared to support and strengthen this hypothesis. The embryonal hypothesis, promoted primarily by Cohnheim, suggested that cancers developed from "embryonal rests", primitive cells originating in fetal life and remaining alive in adults. The infectious hypothesis grew out of the 19th century information explosion on infectious diseases in humans and animals. The popularity of the infectious hypothesis continued into the 20th century, but fell into disrepute in the 1920s because several striking findings, one of which was recognized by the Nobel Prize, were later proved inaccurate. In addition, Yamigawa and Ichikawa demonstrated (Japan, 1914) that chemicals could cause cancer in experimental animals, and this finding attracted considerable interest. Today we know that both infectious agents and chemicals are important causes of human cancer.

Cancer Theories of the 20th Century

With the advent of the "hard" medical sciences of biochemistry, pharmacology, and cell and molecular biology, theories about the origin of cancer were extended from the three basic theories—irritation, embryonic, and infectious—to attempts to understand the molecular mechanisms of cancer. The late Otto Warburg believed that the mechanism underlying cancer development was abnormal cell respiration. He said this resulted in an increase in the fermentation of a sugar (glucose) to lactic acid, a process termed *glycolysis*. The late Jesse Greenstein extended Warburg's concept to propose that the metabolism of all tumors approaches a characteristic pattern. James and Elizabeth Miller of the McArdle Laboratory for Cancer Research suggested almost 40 years ago that neoplasms result from the "deletion" of important proteins and enzymes.

Many other theories have followed the work of these pioneers, but during the last decade our understanding of the basic molecular changes involved in specific types of cancer has suggested that alterations in specific genes within cells may be the critical molecular events in the development of most cancers. At present, the predominant evidence points toward the genetic theory (Figure 2). This theory argues generally that tumor development begins with changes in genetic information encoded in the cell—either by addition, alteration, or subtraction. For example, cancer-causing viruses add genetic information, and chemicals or radiation can alter or subtract genetic information. A portion of the information added by cancer-causing viruses is directly responsible in many instances for the conversion of a normal cell to a cancer cell. These added genes have been given the name "oncogenes". Some of these oncogenes, particularly in RNA cancer-causing viruses, have their counterparts in normal cells; these are termed *proto-oncogenes*. Interestingly, alterations in proto-oncogenes by chemicals or radiation have been described in cancers that resulted from these agents. Thus, it appears that most cancers result from or at least are associated with significant genetic alterations in the cells.

Figure 2. Theories of cancer formation. Genetic: *Information is added to or subtracted from DNA.* Nongenetic: *DNA remains unaltered, but its messages are expressed erroneously.*

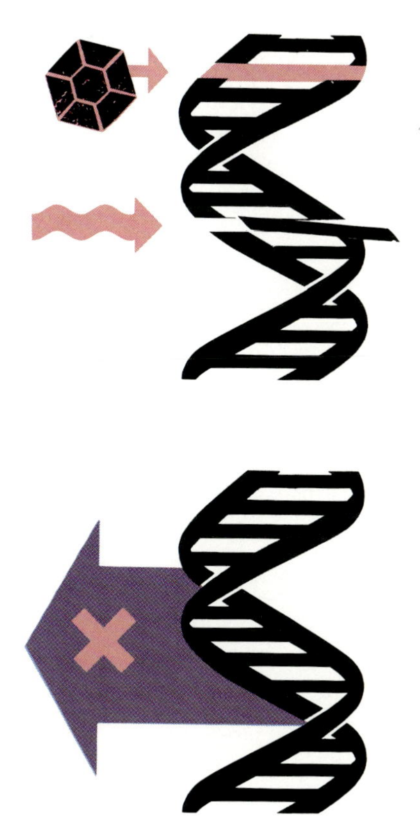

There is also significant evidence in a few instances that tumors can result from nongenetic alterations, comparable to the process of cellular differentiation (Figure 3) during the development of organs and tissues in embryos (Figure 4). Although either the genetic or the nongenetic (non-DNA) concept of the mechanism of cancer formation can account for the hereditary aspect of cancer growth continuing through generations of cells, at present neither can be considered the ubiquitous proven mechanism of cancer formation. It is more likely that both concepts play significant roles at different stages in the development of cancer.

What Is Cancer?

Strangely, physicians did not attempt to reach a meaningful definition of cancer until the 1920s. Since then, several scientists have proposed definitions, usually based on their particular areas of investigation. For this reason, and also because of the variety of cancers, the general definition used here assumes that all varieties have an underlying common, definable theme. This definition is essentially that of the pathologist, J. Ewing, with a slight modification: *A neoplasm is a heritably altered, relatively autonomous growth of tissue.*

Up to now we have been using the term "cancer" or "tumor" in a general way to refer to the disease we know as cancer. The scientific or medical term is "neoplasm". As we shall see later, the term cancer refers to malignant neoplasms. Tumor is a general term indicating any abnormal mass or growth of tissue. Thus, neoplasms are tumors, but so is any swelling or lump, such as a bruise, scar, or mass of repaired tissue around a bone fracture.

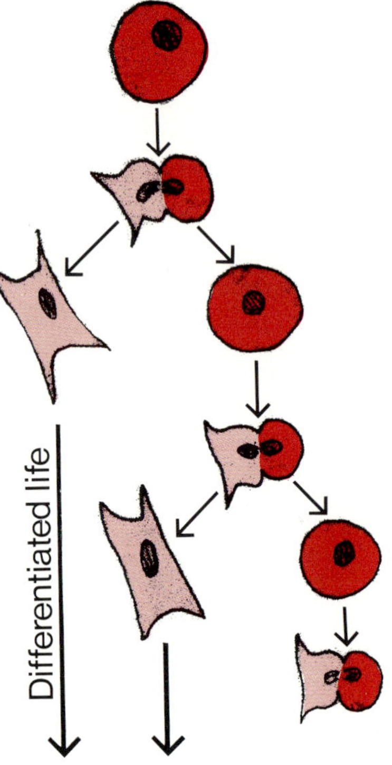

Figure 3. Cellular differentiation. Although the DNA remains unaltered, cells change in appearance and behavior. The new characteristics are passed on to succeeding cell generations.

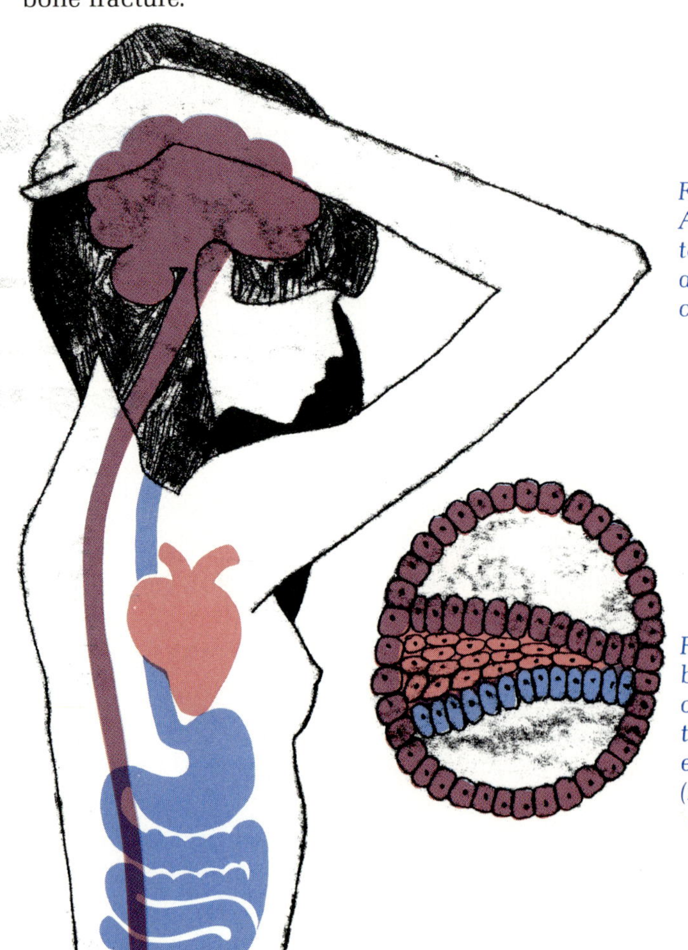

Figure 4. Before the early embryo (lower right) begins to form organs, its cells are arranged in three germ layers, called the ectoderm (purple), mesoderm (red), and endoderm (blue), which develop into specialized tissues.

Ewing's definition implies several concepts. The changes from the normal state exhibited by a neoplastic cell are "heritable" in that such characteristics are inherited by the progeny of the neoplastic cell. "Autonomy" indicates that cancer is not subject to the rules and regulations governing normal cells and the overall function of the organism; this autonomy applies to control of cell division as well as to cell function, such as synthesis of hormones and other cell products. The adverb "relative" indicates that tumors are not completely autonomous. In many instances, the autonomy may be subtle and relative to the tissue from which it arose. The classical example is the relatively uncontrolled production of insulin, a hormone that regulates the level of blood sugar, by a tumor that originates from the islets of Langerhans (the glandular tissue that normally produces insulin). Although the tumor may be no bigger than a pea, it still may induce life-threatening symptoms (Figure 5).

The phrase "relative autonomy" is the most important aspect of the definition; it sets off a particular cell type as neoplastic. The term is used here in the biological sense, but one day we will understand it in the molecular sense; then we will

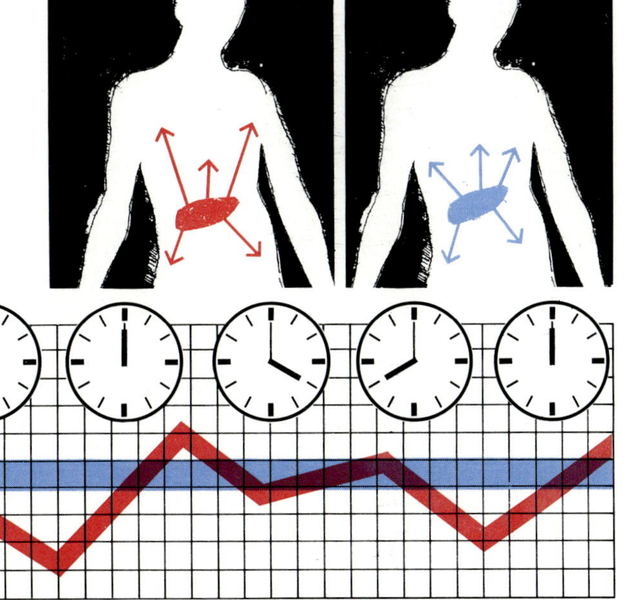

Figure 5. An example of relative cell autonomy. Normally the islets of Langerhans in the pancreas produce insulin only as needed by the body. However, when these cells are cancerous, the cutoff mechanism no longer operates.

probably understand the mechanism of cancer development.

The third component of the definition is the term "growth". Growth may indicate the rate of cell division or the rate at which macromolecules, such as DNA and proteins, are synthesized. The actual growth rate of some tumors is extremely slow, differing little from those of their normal counterparts; in the most serious cases the rate may be extremely rapid, approaching that of embryonic tissue.

On the other hand, tumor cells may exist within the host for its lifetime without ever undergoing demonstrable cell division. The rate of cell replication in some tumors is actually less than that of the cells from which they arose, such as in chronic leukemias and cancers of the small intestine.

Finally, the fourth component of the definition is the term "tissue". This term requires that tumors can be found only in multicellular organisms. By this definition, unicellular organisms are free of this disease. Thus, cancer becomes the curse of evolution.

The biology of cancer observed in living organisms is the basic reference for our definition, but with recent advances in growing cells in the laboratory, the goal has become a molecular definition.

Tumor Classification

Behavioristic. Because the definition of cancer in this chapter is based on biological behavior, a classification based on such behavior is proper (Figure 6). This classification is useful in determining the prognosis of cancer patients, but is of little use to scientists who study mechanisms of tumor formation at the molecular level or to those interested in producing tumors by various agents, including factors in our environment.

Most differences between benign and malignant tumors are relative. The critical difference is

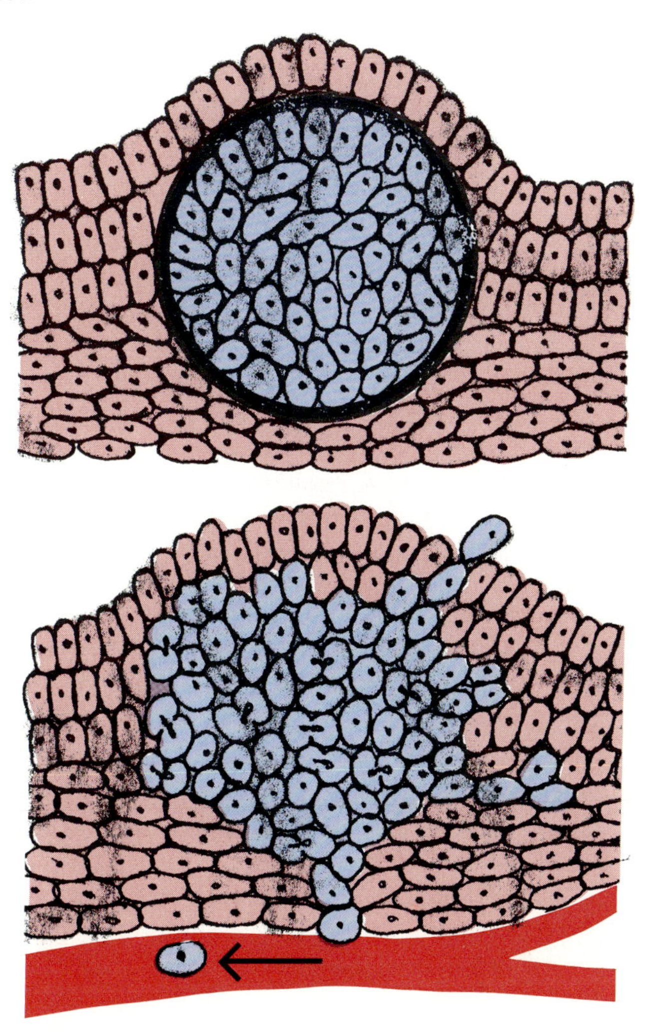

Figure 6. Tumor characteristics. Top, right: benign tumors are encapsulated, grow slowly, and do not invade surrounding tissue, that is, metastasize. Right bottom: malignant tumors grow rapidly, are not encapsulated, invade surrounding tissue, and metastasize. Other characteristics of tumors are

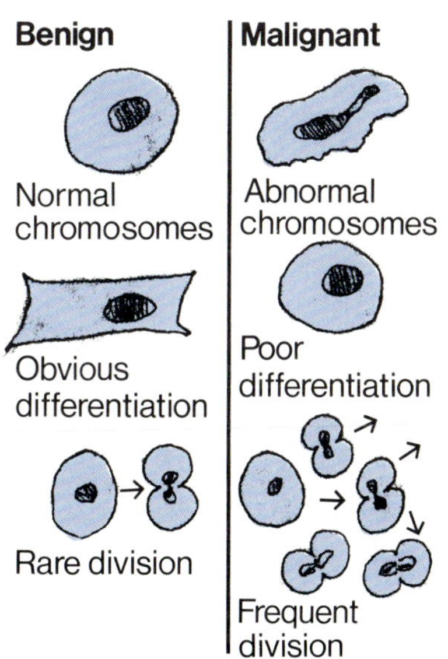

that benign tumors do not metastasize, whereas malignant tumors (or cancers) do. A metastasis is a secondary growth originating from the primary tumor and growing elsewhere in the body.

Other Classifications. Although the behavioristic classification of tumors is commonly accepted, many pathologists prefer a classification based on the tissue of origin. Such classifications distinguish between tumors of tissues such as epithelium, connective tissue, the nervous system, the blood-forming system, and multiple-tissue (mixed-cell) origins.

In classifying or diagnosing a specific tumor, other systems are often used. For example, terms such as papillary, cystic, or follicular may relate to tumors from a variety of cellular origins. In addition, some tumors have been named according to the individual who first described the condition. Examples are Ewing's tumor of bone, Paget's disease, and Hodgkin's disease.

Embryologic Basis of Nomenclature

No system of nomenclature of neoplasms is accepted worldwide. In this country the nomenclature has revolved around the suffix, "-oma", which literally means tumor. With few exceptions, words with this suffix refer to neoplasms. Granuloma, a growth of inflammatory tissue, and hematoma, a mass of blood outside the vessels in a tissue, are exceptions.

Benign tumors are named with a prefix that refers to the tissue from which they arose, together with the suffix -oma. For example, a benign tumor of fibrous tissue is called a fibroma; of cartilage, a chondroma; and of glandular tissue, an adenoma.

Additionally, cancers are divided into two general categories, depending on their embryologic origin. In the early embryo, before organs begin to form, cells arrange themselves in three layers—ectodermal, mesodermal, and endodermal (see Figure 4). Ectodermal cells form skin, its appendages, and nerve tissue; mesodermal cells form bone, muscle, cartilage, and related tissues; and endodermal cells form the intestinal system and its associated organs. A cancer that arises from mesodermal tissue is called a sarcoma. If it arises from embryonic ectodermal or endodermal tissue, it is a carcinoma.

By combining this nomenclature with the tissue from which the cancer arose, one may have an adenocarcinoma (adeno-, glandular; carcinoma, arising from endodermal tissue) of the stomach, pancreas, or breast. On the other hand, one may have a chondrosarcoma of cartilage (chondro-, cartilage; sarcoma, arising from mesodermal tissue), a fibrosarcoma from fibrous tissue, or an osteogenic sarcoma (osteo-, bone; genic, producing) derived from bone.

Unfortunately, all cancer nomenclature does not fit neatly into this pattern. The suffix "-blastoma" is used to indicate certain types of tumors that have a primitive appearance resembling embryonic structures. Examples are the neuroblastoma (neuro-, nerve) and the myoblastoma (myo-, muscle).

Sometimes nomenclature is rather confusing. A highly malignant tumor with the appearance of both a carcinoma and a sarcoma is termed a carcinosarcoma. This term would indicate that the tumor was derived from two embryonic layers. Another cancer, called a mixed tumor of the salivary gland, was thought also to have originated from two embryologic layers; now, however, this mixed tumor is believed to be a slowly growing carcinoma. The most common tumor of multiple-tissue origin is the teratoma, derived from all three embryonic layers. It may be either benign or malignant.

Development of Cancer in the Body

Latency. One of the most common characteristics of the development of a neoplasm in an organism is the extended period of time between the initial application of a carcinogenic (cancer-causing) agent—be it physical, chemical, or biological—and the appearance of a neoplasm. This latency phenomenon or tumor induction time, which can be demonstrated most readily after treatment with chemical carcinogens, occurs even when the carcinogen is administered continuously to an experimental animal. In most systems studied, there is little or no evidence of truly neoplastic growth through much of this latency period. Similarly, a latency period can be demonstrated when the carcinogen is administered to a pregnant animal, as in the case of diethylstilbestrol; here the neoplasms appear much later in the offspring. The latency phenomenon may also be seen after infection with oncogenic (cancer-causing) viruses, ionizing radiation, or in the enigmatic production of sarcomas by the implantation of plastic or metal disks under the skin of an animal. The latency period varies with the type of carcinogenic agent, its dosage, and certain characteristics of the target cells within the host.

Stages in the Development of Neoplasia. Our understanding of the basis for latency in carcinogenesis began more than 40 years ago with the demonstration that a single application of a chemical carcinogen to the skin of a mouse had no effect unless a second agent or process was applied chronically to the site of carcinogen application. It was necessary for both conditions to be met, and the order had to be the application of a carcinogen followed by the chronic administration of the particular noncarcinogenic agent. Even when there was a delay of as much as a year between the first process and the second, the combination still resulted in 100% tumor incidence.

The two processes were termed "initiation" and "promotion" by these early investigators. They found that initiation was irreversible and additive in a dose-responsive manner, and that no threshold, or no-effect dose, could be accurately determined. Promotion, however, was found to be reversible. Withdrawal of the promoting agent well before tumors appeared delayed or prevented the appearance of neoplasms. In addition, changing the dosage schedule of the promoting agent to allow long intervals of time between its applications also resulted in no neoplasms. A no-effect dose could usually be determined, as might be expected from the reversible nature of the process. Furthermore, tumor promotion can be modulated by a variety of environmental factors, including diet, age, hormonal balance, sex, and others.

For many years the stages of initiation and promotion in the development of cancer were studied most thoroughly in the skin, but during the last decade these and the final stage in the development of cancer have been observed in several other tissues. One example that has been investigated widely is the development of liver cancer in rodents. The format for such an experiment, which is entirely analogous to that used with skin cancer, is shown in Figure 7. A single dose of an initiating agent is given to a very young animal, usually within a few days after birth, and administration of the promoting agent begins at weaning. As an alternative, initiation may be accomplished in the

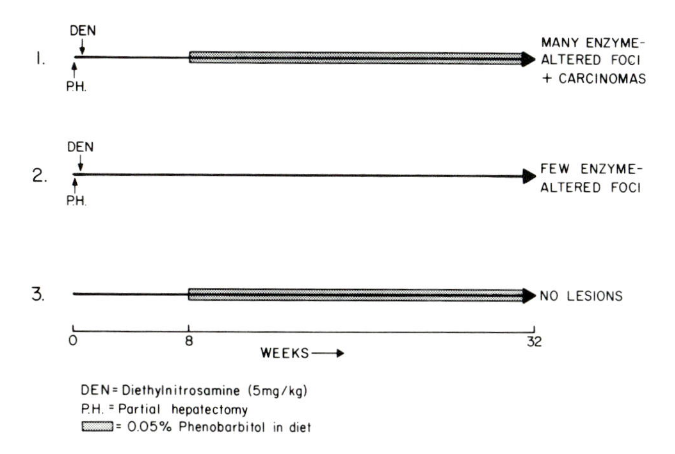

Figure 7. Schematic representation of the protocol illustrative of the stages of initiation and promotion during carcinogenesis in the liver. Initiation is accomplished with a single dose of diethylnitrosamine (DEN) given when liver cells are dividing, either shortly after birth or in adult life following partial surgical removal of the liver. Eight weeks later some groups are placed on a diet containing the promoting agent (phenobarbital).

adult by chemical or surgical stimulation of cell division in the liver.

It is clear that initiation requires cell division for the "fixation" of the heritable change induced by the initiating agent, much like the fixation requirement for a chemical mutation in cells. Under these conditions, about 1 in 10^5 cells becomes initiated, out of 10^9 total cells in the liver. However, 10,000 cancers do not result from this large number of initiated cells. During the period of tumor promotion, each initiated cell divides many times and forms a small colony of cells. The promoting agent acts to stimulate the increased replication of the cells of each of these colonies or foci. If, however, the promoting agent is removed, the cells within most of the foci stop replicating and either die or, in certain situations, apparently become essentially normal cells. Readministration of the promoting agent allows the foci to develop again, presumably from a few cells that survive in the absence of the promoting agent. Thus, the progeny of initiated cells depend on the presence of the promoting agent for their growth and special characteristics.

With such experimental systems, it is possible to characterize specific chemicals as either initiating agents (also termed incomplete carcinogens), promoting agents, or complete carcinogens (having the capability both to initiate and to promote neoplastic cells). Few true incomplete carcinogens are known, but many complete carcinogens have been identified, including cigarette smoke, aflatoxin (a natural mold toxin that contaminates some foodstuffs), many cancer-causing viruses, and ionizing radiation. Examples of promoting agents include phenobarbital, many hormones, dioxin, and tetradecanoylphorbol acetate, a complex skin promoter that has been studied extensively.

Several conditions in humans exhibit characteristics analogous to the stages of initiation and promotion seen in animals. Individuals who stop smoking decrease their risk of developing lung cancer for each year of disease-free survival after quitting. This effect is to be expected, as it interrupts the reversible promotion stage of carcinogenesis caused by cigarette smoking. Some human neoplasms show structural characteristics of malignancy, such as increased rate of cell division and abnormal appearance when viewed under the microscope, but are benign. When these neoplasms occur in epithelium (such as skin, intestinal surface, or glands), they are called carcinomas in situ. The most common example is found in the human cervix, the entrance to the uterus. Such microscopic focal alterations normally occur in women between ages 30 and 40, whereas invasive malignant cervical carcinoma usually occurs after the age of 40. This relationship suggests that the focal carcinoma in situ represents recently initiated cells, whereas the ultimate malignancy represents stages beyond this.

Tumor Progression. Although early investigators suggested that the stage of promotion extended to the appearance of malignant neoplasms, the reversible nature of this stage did not conform to the irreversible nature of cancer itself. In recent years it has become apparent, especially in the experimental systems in skin and liver, that a third stage, now termed progression, follows the reversible promotion stage (Figure 8). Progression is characterized by demonstrable changes in the neoplastic cells associated with increased growth rate, increased invasiveness, metastases, and alterations in biochemical and structural characteristics of the neoplasm.

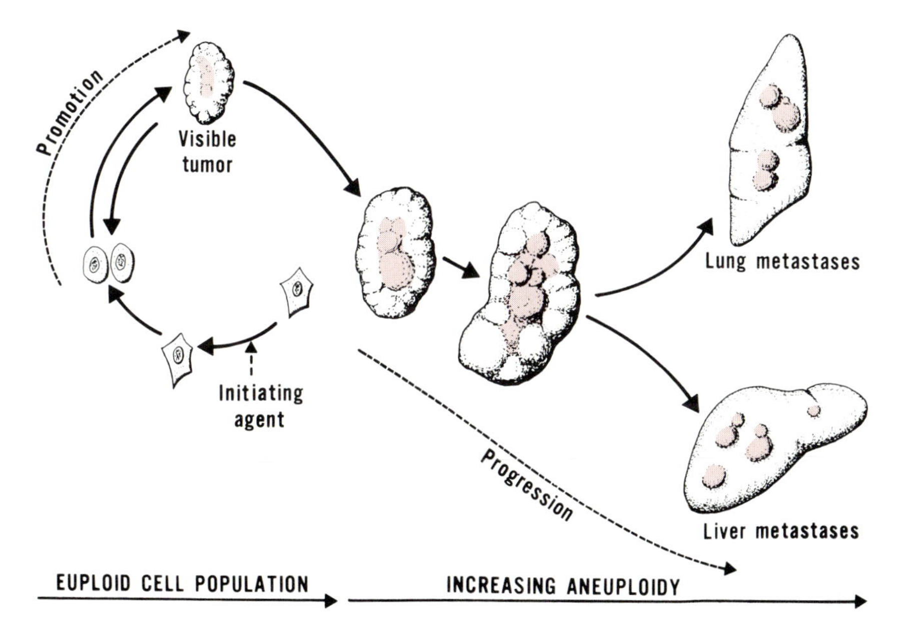

Figure 8. The natural history of neoplastic development, beginning with the initiated cell resulting from the administration of an initiating agent (carcinogen), followed by promotion to a visible tumor and progression of this neoplasm to the malignant or cancerous state. Euploid and aneuploidy refer to the characteristics of the chromosomes within the cells, the former indicating a normal complement of chromosomes, while aneuploidy denotes abnormalities in number or structure of one or more chromosomes (adapted from Pitot, H. C., Fundamentals of Oncology, 3rd ed., Marcel Dekker, 1986, *with permission of the publisher).*

These alterations correlate with changes in the number or arrangement of genes (as evidenced from nucleic acid sequence and hybridization studies) or with visible chromosomal alterations (as evidenced by light microscopic techniques) within a majority of the neoplastic cells that make up the tumor. Clear, heritable abnormalities in the cell genome can be readily demonstrated in this stage. Furthermore, neoplastic cells in the progression stage are characterized by an instability of their karyotype, the morphology and number of their chromosomes. Such instability leads to an increase in specific genes or sets of genes, or to rearrangements of genes in such a way that factors controlling gene expression change dramatically. In this way, increased expression of proto-oncogenes may occur during progression; this often occurs in cancers induced by chemicals or radiation in both animals and humans. Such increased expression of proto-oncogenes has not been shown in the stages of initiation and promotion, although mutations in proto-oncogenes have been found during these early stages. Karyotypic alterations appear to enhance the ability of a cancer to metastasize. As will be shown in a later chapter, such metastases demonstrate karyotype instability and heterogeneity of their genomic structures. In all likelihood, cells with certain abnormal karyotypes are best suited for metastatic growth.

Tumor Regression. Although it is not usually considered a stage in the development of neoplasia, we know now that both benign and malignant tumors may regress or, more specifically, differentiate. In some instances normal cells or their products appear to exert an influence on the regression and differentiation of malignant neoplasms, quite beyond the obvious mechanism in which the host recognizes the tumor as foreign tissue and rejects it immunologically. Transplantation of malignant cells into early embryos has resulted in the development of such malignant cells to normal cells.

The chemical induction of differentiation of malignant neoplasms was recently demonstrated in vitro. Thus, it is possible that some or all neoplasms exhibit some degree of differentiation and regression of characteristics. The best human example of this is neuroblastoma, a malignant neoplasm seen primarily in children. Under some circumstances it may differentiate into masses of benign, normal-appearing nerve cells. Unfortunately, we have not yet been able to reproduce this process in vivo in a controlled manner. However, the potential for regulating neoplastic growth by the forced regression or differentiation of cancer cells is now apparent.

Conclusion

This introduction to the biology of neoplasia will serve as a foundation for the more detailed concepts presented in the following chapters. The information presented here is intended to encourage the reader to think seriously about what we can do to prevent, understand, and ultimately to eradicate cancer in our society.

Suggested Reading

Bett, W. R. "Historical Aspects of Cancer." In *Cancer, Vol. 1;* R. W. Raven, Ed.; Butterworth: London, 1957; pp 1–5.

Ewing, J. *Neoplastic Diseases: A Treatise on Tumors;* W. B. Saunders: Philadelphia, 1940; 4th ed.

Freshney, R. I. "Induction of Differentiation in Neoplastic Cells." *Anti-cancer Research,* **1985,** 5, 111.

Granberg, I. "Chromosomes in Preinvasive, Microinvasive and Invasive Cervical Carcinoma." *Hereditas,* **1971,** *68,* 165.

Greenstein, J. P. "Some Biochemical Characteristics of Morphologically Separate Cancers." *Cancer Res.,* **1956,** *16,* 641.

Pitot, H. C. *Fundamentals of Oncology;* Marcel Dekker: New York, 1986; 3rd ed.

Pitot, H. C. "Oncogenes and Human Neoplasia." *Clinics in Laboratory Medicine,* **1986,** 6, 167.

Pitot, H. C.; Beer, D. G.; Hendrich, S. "Multi-stage Carcinogenesis of the Rat Hepatocyte." In *Nongenotoxic Mechanisms in Carcinogenesis;* Banbury Report 25; Cold Spring Harbor Laboratory: Cold Spring Harbor; pp 41–53.

Pitot, H. C.; Sirica, A. E. "The Stages of Initiation and Promotion in Hepatocarcinogenesis." *Biochim. Biophys. Acta,* **1980,** *605,* 191.

Ritchie, A. C. "The Classification, Morphology and Behavior of Tumors." In *General Pathology;* Florey, H., Ed.; W. B. Saunders: Philadelphia, 1962; pp 551–597.

Sachs, L. "Growth, Differentiation and the Reversal of Malignancy." *Scientific American*, **1986,** *254*, 30.
Warburg, O. "On the Origin of Cancer Cells." *Science (Washington, DC)*, **1956,** *123*, 309.
Wilson, L. M. K.; Draper, G. J. "Neuroblastoma, Its Natural History and Prognosis: A Study of 487 Cases." *Brit. Med. J.*, **1974,** *8*, 301–307.
Report of the National Panel of Consultants on The Conquest of Cancer; U.S. Government Printing Office: Washington, D.C., 1971.

CHAPTER 2 Cell Growth Control and Cancer

Arthur B. Pardee and James G. Rheinwald

The human body contains more than 10 million million cells. Many of them divide to produce new cells; these cells, in turn, grow and divide again. Some cells go through a process called differentiation, in which they stop dividing, develop specialized structures and functions, and may eventually die and be replaced by the division of other cells. Cell division and differentiation are guided by natural control mechanisms. These occasionally break down, and the result is cancer. Scientists are beginning to gain an understanding of these control processes and what has gone wrong with them in cancer cells, but they need to know why cancer cells grow at times, rates, and places that they should not. In trying to understand cancer, scientists must compare the cancer cell with the normal cell from which it originated. Thus, they may consider

- how cancer tissues and cells differ from normal tissues and cells
- laboratory techniques for studying cell growth
- factors and mechanisms regulating growth and differentiation
- the biochemistry of cell growth and the duplication of a cell's components during its lifetime

1420–4/88/0019$06.00/0

- the application of this knowledge about normal growth control to the problem of cancer

Comparing Cancer with Normal Tissues

What is cancer? Like all tissues, cancers are groups of cells that arise from a single cell. But cancer cells are different in that they grow abnormally, both by dividing excessively to form lumps or tumors and by failing to differentiate normally. Groups of cancer cells look different from normal cells under the microscope. Pathologists can detect cancer by the abnormal appearance of the cells in slices taken from malignant tissues. Fundamentally, however, the defect lies invisibly within each cancer cell. Thus we consider cancer a cellular disease, not a disease of organs.

Normal cells sense signals from hormones and other chemicals circulating in the body and from contacts with neighboring cells. They respond correctly by growth and differentiation that is appropriate relative to the rest of the body. What causes a single cell to change so that it ignores these environmental signals heeded by normal cells?

We can think of a human being as an organized cluster of roughly 100 different cell types. Each cell type grows at a certain rate and in a certain way and place during the human life cycle. What are some of the normal growth patterns for cells? Brain cells seldom or never divide in adults, though fortunately they function well for many years in such a differentiated state. Many cell types normally remain in a nondividing "resting" state, but can respond quickly to a signal and reproduce as needed. A familiar example occurs whenever we happen to cut ourselves. Almost immediately, nearby dermal cells called *fibroblasts* begin to divide and make new cells to fill the wound. After repair, these cells stop growing and the wound smooths over. Hence, resting fibroblasts retain the ability to divide again. When our body signals its need for more cells, more are produced.

Other types, such as blood cells, skin epidermal cells, and the cells that line the intestine, divide very frequently to produce differentiated cells that serve their function for only a limited time before they deteriorate and must be replaced. Consider more closely our red blood cells. In humans they survive about three months and then die. They are continually being replaced by the vigorous division of *stem cells* in the bone marrow. As the body uses up its new blood cells, these stem cells remain ready to divide again. This process is remarkably balanced; new cell production precisely equals cell loss.

In cancer a surplus of new cells results from blatant disregard of this balance mechanism. For example, in leukemia, which is a cancer involving blood-producing tissues, an excess of white blood cells (also called leukocytes) is formed. Leukemia cells do not await environmental control signals such as those generated when the circulatory system calls for more blood cells. Instead, they multiply continuously and excessively.

Cancer cells have other important properties besides their loss of growth regulation. They also are defective in differentiation. Sometimes they invade surrounding tissues, and thus disturb normal tissue structures and functions. A pathologist often sees cancer cells mixed in disarray with normal cells. Cancer cells can also break away from their starting point and wander through the blood or lymph fluid, to settle elsewhere and grow into secondary tumors. This process, called *metastasis*, poses a severe problem in treating cancer. Even if the original tumor is discovered and completely removed by surgery or X-rays, many secondary metastatic tumors may have already formed and eventually could kill the patient. Many cancers grow slowly, and yet they too can be lethal if they damage normal tissues and are too widely spread to be removed.

Much evidence shows that cancer develops by a multistep process. Loss of growth control occurs gradually, and normal controls are only slightly relaxed at first. Such premalignant cells later produce new ones that become less growth restricted and gain properties such as invasiveness and metastasis. Some agents such as chemicals or radiation can cause the first step in cancer, called initiation. Then other agents called promoters can make the initiated cells progress to increased malignancy.

Techniques for Studying Cancer in the Laboratory

Studies of the basic nature of cancer are difficult to perform with humans and even with animals, because scientists cannot easily change the environment in which cells grow in vivo, that is, inside the animal. Growing cells outside of an animal creates more readily controlled conditions. Fortunately, important properties of cells remain intact when they are maintained as cultures in the laboratory. These properties include the ability to respond to growth and differentiation signals and to drugs in much the same way as cells do in the animal.

The technique for growing isolated cells (Figure 1) in the laboratory is called *tissue culture*. A small tissue sample is removed from an animal or human. The cells are separated from one another, placed into a glass or plastic container, and covered with a nutrient solution called a growth medium. The culture is usually incubated at 37 °C, the temperature of the body. Although cells will not grow in culture exactly as they do in the body, their growth is orderly and mimics that of cells in vivo in many ways. For example, they remain responsive to chemical *growth factors* added to the culture medium and they stop growing when they become crowded.

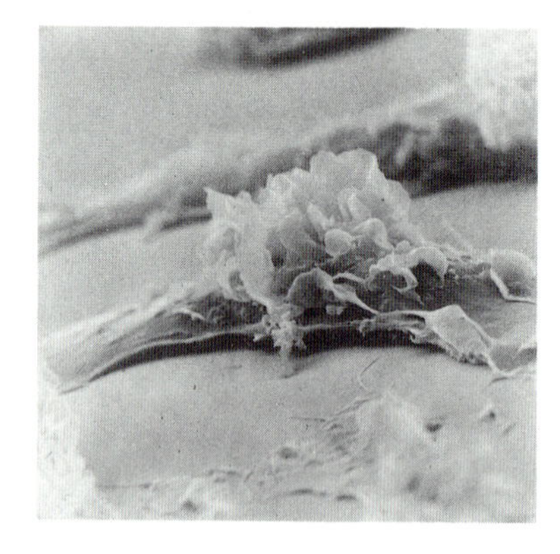

Figure 1. Scanning electron microscope picture of an isolated cell.

Important differences have been found in the growth properties of normal and cancer cells. We will discuss the ways in which cancer cells in culture show their defective growth control. Furthermore, normal cells can be made to undergo *transformation* in culture, which means that they gain some or all of the properties of cancer cells. For example, certain viruses that cause cancer in mice

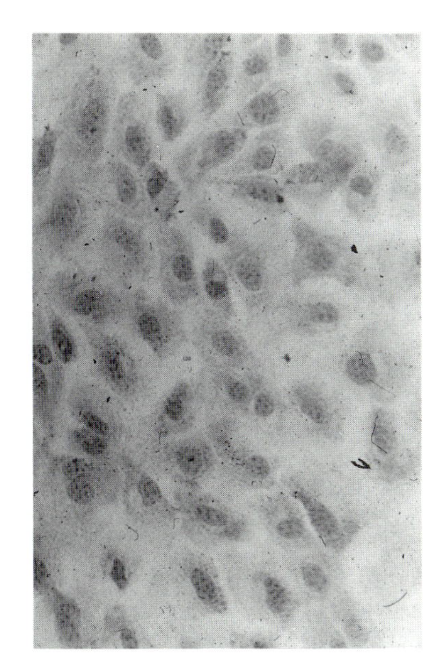

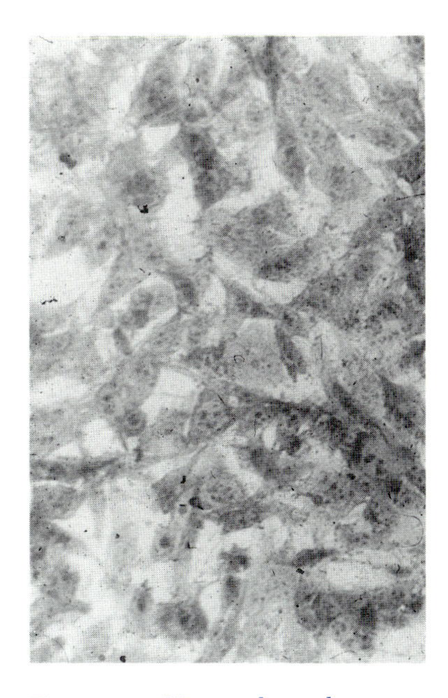

Figure 2. Normal and tumor cells on a surface show distinctly different growth patterns. (Reproduced with permission from Pardee and Prem veer Reddy. Copyright 1986, Carolina Biological Supply Company.)

can also transform normal mouse cells in tissue culture. Soon after virus infection, the cell begins replicating its DNA (the genetic material) independently of external growth factors. These transformed cells also show other growth and biochemical characteristics of cancer cells. Scientists can therefore investigate *carcinogenesis*, the origins of cancer, with cells in culture. Also, chemotherapy—the selective killing of cancer versus normal cells—can be studied in culture, and such studies help develop new methods of chemotherapy for patients.

Factors That Regulate Cell Growth

Cell growth in tissue culture is affected by several of the same conditions that affect cell growth in the animal. One is attachment to a surface. Most normal cell types in the animal are attached to a structured network secreted by the cells, called the extracellular matrix. These cells usually need to attach to a surface such as plastic or glass in order to grow in culture (Figure 2). In contrast, many cancer cells can grow in culture even if they are not attached to a surface.

Another property that affects growth is cell density. Normal cells stop growing when they become overcrowded. After a few cells are added onto the surface of the culture vessel, they settle and begin growing. When each cell has doubled its mass it divides into two daughter cells. This continues only so long as the culture vessel's surface is free to accept new cells. For example, once fibroblasts have formed a layer one cell thick that covers the surface, they stop growing. In a classic experiment, a "wound" can be made in a single layer of cultured fibroblasts by scraping some cells away from the bottom of the culture vessel. The remaining cells at the edge of the wound divide to fill the empty space and then stop when it is filled. Thus, fibroblasts in tissue culture respond to one of the growth controls seen in living animals.

This density-dependent growth inhibition indicates that normal cells somehow recognize and adjust their growth to their neighbors and the available space. In contrast, cancerous fibroblasts, *sarcoma* cells, can continue growing even after they form a monolayer; they then grow on top of one another. Thus cancer cells in culture, like those in the animal, have lost the property of growth inhibition in response to high cell density.

Another factor influencing normal growth is the availability of nutrients and growth-stimulating chemicals provided in the culture medium. Cells themselves cannot synthesize either their food or certain complex building blocks that they need, such as amino acids and vitamins. If any one of these is insufficiently supplied, cells stop growing. Cells in culture also require growth factors, including small proteins such as insulin, epidermal growth factor (EGF), and platelet-derived growth factor (PDGF) in order to grow. Some of these growth factors can be provided easily by giving the cells serum—the clear fluid that remains after blood has clotted and its cells have been removed. Other factors must be purified from animal organs. When the specific growth factors required by a particular cell type are not provided, that cell will not grow in culture. Interestingly, cancer cells almost always have much less need for these growth factors than do normal cells.

Cells in the living animal also require these nutrients and growth factors. Cancer cells in an animal will grow to form a solid tumor mass no larger than a pinhead (one millimeter in diameter) unless they acquire a good blood supply to provide hormones, nutrients, and oxygen. Tumors can induce the growth of new blood vessels into themselves, a process called angiogenesis, which provides them with these factors.

Thus we see that external factors—availability of growth factors and nutrients, of space, and absence of crowding by nearby cells—govern the growth of cells in the animal and in culture. Tumor cells show a decreased sensitivity to all these growth-limiting conditions.

Role of the Cell Surface

External factors control growth by modifying events within the cell. To transmit their external signals into the cell, growth factors act by combining with specific receptors located on the cell surface. For example, epidermal growth factor (EGF) is a small protein of 55 amino acids (building blocks of proteins). Because its structure is known in great detail, Figure 3 is more complicated than the other information presented in this chapter. It is included as a reminder of the depth and detail of knowledge underlying the apparently simple summary of research on growth regulation presented here. Scientists have progressed over the course of 20 years from the initial observation that injection of a crude tissue extract caused a baby mouse's eyelids to open earlier than normal to the detailed molecular and genetic structure of EGF (the responsible substance) and of its receptor.

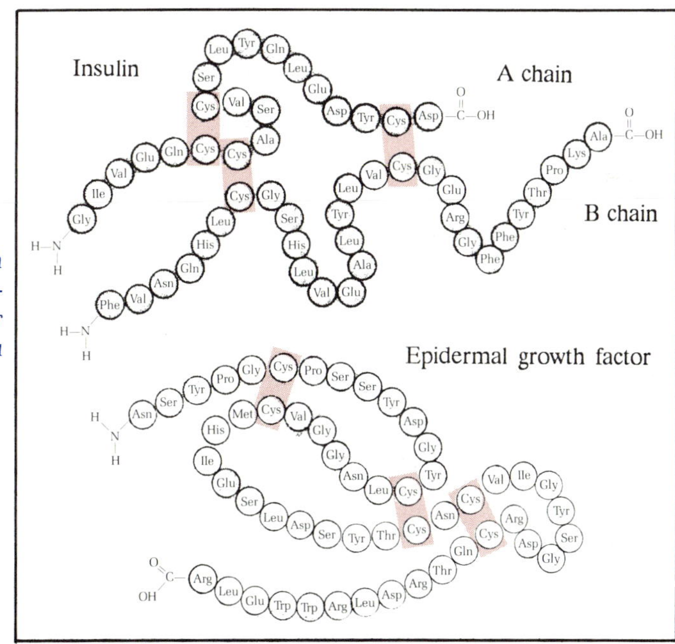

Figure 3. Formula of a growth factor. (Reproduced with permission from Pardee and Prem veer Reddy. Copyright 1986, Carolina Biological Supply Company.)

EGF is recognized by the EGF receptor, a well-defined large protein that is present at about 100,000 copies per cell. This receptor posesses regional structures that permit it to carry information from the cell's outside to its interior. The EGF receptor protein extends across the cell membrane; one end is on the outside and the other on the inside. The external part forms a specific and very strong binding site for EGF. A short region in the middle spans the cell membrane. Another large part of the receptor is exposed to the inside of the cell. This is an enzyme, a catalyst for a chemical reaction. When EGF binds to the receptor's outside part, its structure is changed so that its enzyme portion inside the cell becomes active. This enzyme, a protein kinase, catalyzes the addition of phosphate to some proteins inside the cell. In this way the cell passes information about what is outside itself to its internal biochemical machinery in order to regulate growth.

EGF is an essential factor for most epithelial cell types. Very interestingly, most tumors derived from epithelial cells (called *carcinomas*) have a greatly reduced requirement for EGF. Carcinoma cells bypass the normal requirement of EGF for growth and so continue to grow even when EGF is in short supply. Some kinds of tumor cells produce a much larger number of EGF receptors, so that a small amount of EGF generates a more potent signal. Another bypass mechanism is for the tumor cell itself to produce an EGF-like factor, called tumor growth factor alpha (α). Such cells constantly bathe themselves with this factor, which stimulates their own EGF receptors. Thus they are constantly turned on for growth. A third mechanism is an internal short-circuiting of the whole EGF receptor mechanism we have just described. Some signal that normally takes place after EGF binding to its receptor is constantly present in these cells, and so they behave as if EGF is always bound. Clearly, cancer cells have found more than one way to evade the growth factor requirements that keep normal cells under control.

Intracellular Events in Growth Regulation

Once growth signals have crossed the cell membrane via receptors, their information needs to be passed on to the nucleus. Various processes that are performed in the nucleus are needed to start duplication of all of the thousands of different molecules that will form the new cell.

Just how the signal is passed from the stimulated receptor kinase on the inner side of the cell membrane all the way into the nucleus is not yet fully understood and is under vigorous investigation. "Second messengers" are proposed—molecules that are activated or produced by the growth factor receptor and then move to the nucleus (Figure 4). Among a number of candidates for such second messengers are proteins that have gained new phosphate groups added by kinase enzymes such as the kinase of the EGF receptor. Phosphorylation can give a protein new properties, including a change in its enzymatic activity. Several internal proteins have been found to be newly phosphorylated following the addition of growth factors.

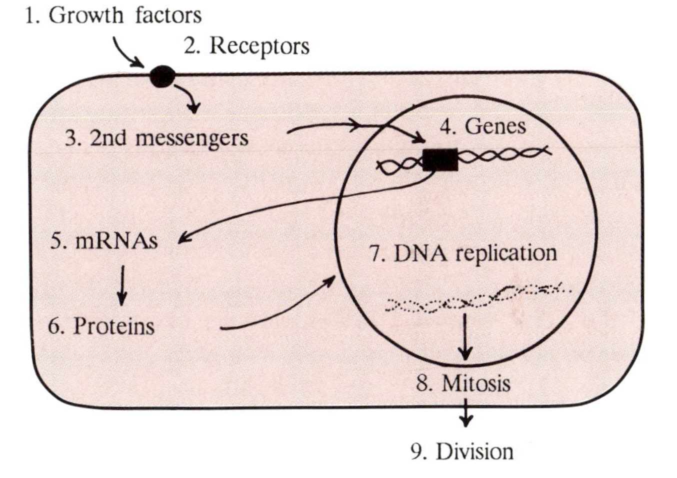

Figure 4. Events in cell proliferation. (Reproduced with permission from Pardee. Copyright 1985, Allan R. Liss, Inc.)

Molecules much smaller than proteins constitute another class of second messengers. Some of them are very simple. For example, the concentrations of calcium ions, sodium ions, and even hydrogen ions inside the cell become altered by growth factors. Concentrations of some small organic molecules also change rapidly and dramatically; among these are cyclic AMP and inositol phosphates. The complex, enzyme-catalyzed reactions that produce and destroy these molecules are quite well understood, but are beyond the scope of this chapter. However, the processes by which they act to bring about cell division are still largely unknown.

Gene Activation

One of the most exciting research topics in biology today is the effort to understand the final steps by which hormones and growth factors turn on genes. Genes are composed of long sequences of nucleotide base pairs (building blocks) in the DNA. Some sequences are so-called "open reading frames", which code for the amino acid sequences of proteins. At both ends of these DNA coding regions are sequences important for properly regulating expression of the coding sequences, by turning on or off the copying of DNA into an RNA message that will be used to direct the synthesis of the particular protein.

Gene expression is controlled by the binding of activating or repressing proteins to the regulatory regions. Mutations placed experimentally at important places within these regulatory regions modulate the appearance of the gene's product: messenger RNA (mRNA). "Footprinting" experiments can reveal the protein–DNA interactions that control gene expression. When a gene is turned on (as the ultimate consequence of growth factor binding to the cell), specific DNA sequences become covered by regulatory proteins. These sequences are no longer susceptible to reactive chemicals whose effects on the DNA can be measured. The attachment of regulatory proteins of DNA sequences presumably depends upon the second messages which, in turn, were activated by growth factors interacting with their receptors.

Oncogenes

Another approach to studying the regulation of cell growth and differentiation began with the discovery of special DNA sequences called *oncogenes* in cancer-causing animal viruses and later in human tumor cells. As discussed elsewhere in this book, normal forms of these genes, called *proto-oncogenes*, are present in normal cells. These appear to have roles in normal embryonic development, cell growth control, and differentiation. One or more of these proto-oncogenes are often found to be mutated in cancer cells. These mutations affect either the regulation or the function of the protein that the proto-oncogene encodes. This change results in abnormal growth control. Actions of the various oncogenes in cancer cells are diverse; some cause the production of growth factors, and others modulate receptors for these factors. Some oncogenes encode protein kinases, which perhaps substitute for growth-factor-regulated kinases of normal cells. The products of still other oncogenes are located in the nucleus, where they may alter the expression of genes important for cell division.

The stimulation of cell replication, whether in a normal or cancer cell, involves the activation of genes that are less active or are unexpressed in the resting cell. An activated gene produces mRNA molecules, which are copies of the gene's coding sequences plus parts of its adjacent upstream and downstream sequences. New techniques of molecular biology allow these mRNAs to be detected and isolated. New mRNAs appear when cell growth is activated, but the functions of most of these are not yet known.

The next stage in growth stimulation is "translation" of these messenger RNAs, that is, the synthesis of new proteins from them. Total protein synthesis increases, as would be expected, when a cell has to duplicate all of its proteins as well as its other molecules. A relatively small number are identified as new proteins, that is, ones not present in resting cells. The functions of these new proteins are not known. Most of them are present in both growing normal cells and cancer cells. This late biochemical step of the growth-stimulatory pathway in normal and cancer cells, production of the essential machinery to start DNA replication, is an important area of cancer research.

Growth and the Cell Cycle

Cell division is the visible end of the process that started with growth factor stimulation. One cell becomes two. A cell cycle is the period of time from the birth of a new cell to the time it divides. Myriads of events occur, many of which take place at a definite time preceding cell division (Figure 5). Most important is the duplication of DNA, both in the cell's nucleus and in one of its organelles, called the mitochondria. DNA replication must be performed with great precision to provide each daughter cell with a complete and accurate version of the organism's genetic information.

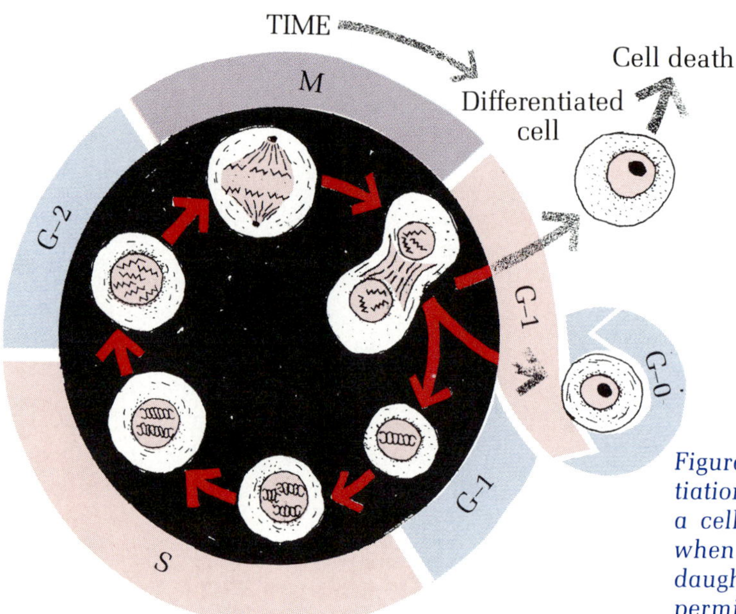

Figure 5. Cell cycle and differentiation. A cell cycle begins when a cell is produced and finishes when that cell divides into two daughter cells. (Reproduced with permission from Pardee. Copyright 1986, Carolina Biological Supply Company.)

A cell's nuclear DNA replicates during only one specific part of the cell cycle—a phase called S, for synthesis. Many mammalian cells growing in culture divide about once every 18 to 24 hours; their S phase occupies about one-third of this total time period. Between the previous cell division and S is a period called G_1 (meaning gap one), during which cells increase in volume but do not make DNA. Many other molecules are synthesized in G_1, including enzymes, structural components, and molecules later involved in the synthesis of DNA. Another segment, G_2 (gap two), exists between the end of S and the phase called mitosis (M). During M the two DNA copies separate and migrate on a spindle apparatus made of protein filaments that distributes them into the two daughter cells. These four main phases of the cell cycle (G_1, S, G_2, and M), all occurring in proper order, are necessary for new cell production.

A growing cell, whether it is normal or cancerous, can pass through the cycle in the same fairly short time. A difference is in special mechanisms that decide (in G_1) whether a cell shall grow or whether it shall enter a resting state, often called G_0. Growth factors in the G_1 part of the cycle act on the switching mechanism that determines whether or not a cell will grow. Cancer cells have defects in this growth-control switch.

Cell Differentiation and Cancer

A second aspect of cell behavior, interconnected with cell cycle activities, is important in our consideration of cancer. This aspect is differentiation. The term has three distinct meanings to a biologist. It can mean the developmental process by which a primitive cell of the early embryo becomes committed to become a single, specialized cell type in the mature animal. It can mean the expression of unique proteins and functions by a particular cell type. Lastly, and most important to our understanding of the cancer cell, differentiation can mean the permanent departure of a cell from the cell cycle, for the purpose of better serving its specialized function. This is usually called *terminal differentiation*, because the commitment to stop growing and to differentiate is irreversible. Subsequent changes undergone by the differentiated cell do not permit it to ever "change its mind", so to speak, and begin dividing again. Examples of terminal differentiation are the accumulation of hemoglobin molecules and loss of the nucleus and chromosomes by red blood cells, the fusion of many individual muscle cells to form large muscle fibers, and the accumulation by epidermal skin cells of densely packed arrays of keratin filaments at the expense of cellular organelles and the nucleus.

Some cell types, such as the fibroblast (the spongy, elastic underlayer of the skin) and the hepatocyte (the functional cell of the liver) do not undergo terminal differentiation. Instead, they perform their functions while in the reversibly resting G_0 state. If a wound occurs in the dermis or liver, elevated local levels of growth factors and hormones, along with a lessening of cell density, stimulate the cells to divide until the correct density is again achieved. While the cells are dividing, they stop expressing most of their differentiation-related genes. Instead they express their growth-related genes until they return to the G_0 state.

We mentioned earlier a "switch" mechanism for shuttling a cell between G_0 and G_1 on the basis of the level of growth factors sensed by the cell. Cells of terminally differentiating tissues such as the epidermis and the blood system seem to have an additional switch in G_1 that triggers the commitment to terminal differentiation. The so-called stem cells of these tissues spend most of their time in G_0. When the tissue calls for more of the terminally differentiated cells, the switch to divide is thrown in some of the stem cells to make them pass through the cell cycle. Some of the daughter cells that arise from stem cells are responsive to factors that switch them to terminal differentiation. One terminal differentiation factor that has been identified and well characterized is erythropoietin. This factor stimulates the daughter cells of blood stem cells to divide about ten more times. While dividing, these cells make more and more hemoglobin and other necessary functional proteins before they finally lose their nuclei and become mature red blood cells.

Various chemical agents efficiently trigger differentiation in particular cell types growing in culture. Because these chemicals are not normally present in animals, their activity tells us little about terminal differentiation mechanisms except that differentiation-triggering agents do not instruct the cell in how to differentiate. Instead, they activate a complex program that has been latent in the cell's chromosomes ever since embryogenesis, when the cell had first become, for example, an epidermal stem cell or a blood stem cell. They are like the on–off switch of a programmed computer.

Cancers of blood cells (leukemias) or of skin epidermal cells (carcinomas) sometimes have a defective differentiation switch. These cancer cells have lost their sensitivity to terminal differentiation signals. Alternatively, some cancers are able to express some of the specialized functions normally reserved for terminal nondividing cells. Both sorts divide indefinitely, as would a constantly stimulated stem cell. A possible avenue for cancer chemotherapy is to develop agents that more effectively trigger terminal differentiation in cancer cells. Such agents might be more selective and, in general, less toxic than chemicals that attack both normal and malignant dividing cells.

The Lifespan of Normal Cells and "Immortality" of Cancer Cells

Normal human cells have the ability to divide as many times as is necessary to fulfill the needs of the organism throughout the human lifespan. When human cells placed in culture are continually stimulated to divide, and are given the opportunity to keep dividing by transferring them to new vessels at lower cell density whenever they become crowded, they are only able to undergo about 50 to 150 doublings. The precise number of total doublings depends upon the cell type, the culture medium used, and the characteristics and age of the individual from whom the cells were taken. Little is known about the mechanism that causes this *replicative senescence* in normal cells.

When cancer cells are examined in culture, they are frequently found to have no limit to the number of times they can divide (Figure 6). Thus cancer cells are often *immortal*, besides being less dependent upon growth factors for stimulating each division. When an immortal cancer cell is fused with a normal cell, or two different cancer cells are fused together, the resulting hybrid cell usually, but not always, goes on to divide only a limited number of times and ultimately senesces. One interpretation of this phenomenon is that several genes together encode an active mechanism that counts and eventually limits the number of times a normal cell's genetic material can be replicated. In cancer cells, one or more of these senescence genes may be inactivated, either by mutation or by partial chromosome loss. Normal cells of mice, rats, and hamsters can be selected during growth in culture for mutations in their senescence mechanism, which permit them to continue to grow without limit. Many of these immortal cell lines are not malignant (lethally cancerous) when they are injected into experimental animals, nor are they greatly altered in other important growth control mechanisms, indicating that immortality itself does not make a cell a cancer cell.

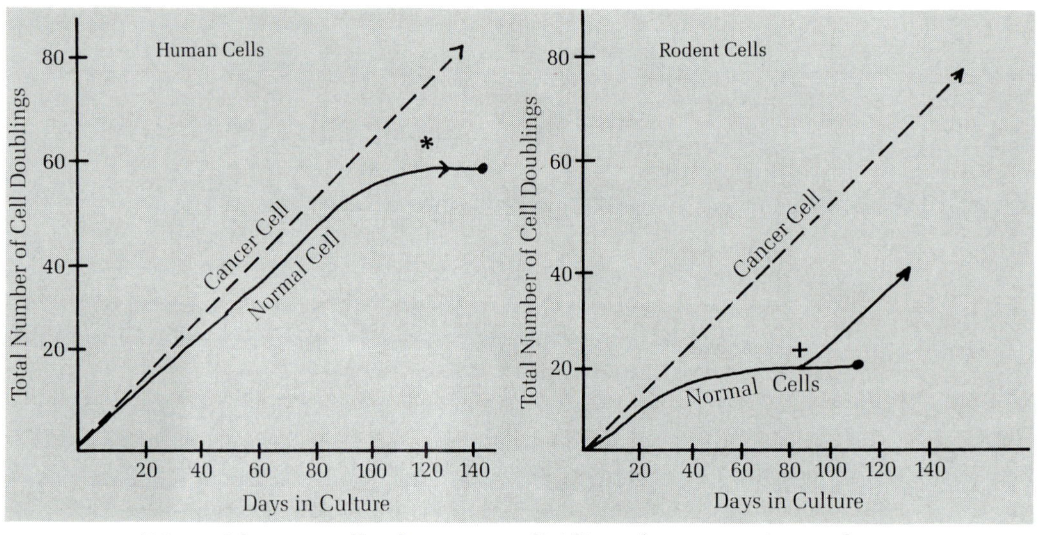

*Normal human cells always stop dividing after a certain number of doublings in culture.

+Normal rodent cells usually become senescent, but some cells undergo genetic changes such that they are able to divide indefinitely. These "immortal" cells are often not cancerous, but are considered "premalignant" because they readily undergo further changes to become malignant.

Cancer cells from either species usually are able to divide without limit.

Figure 6. Senescence vs. immortality: differences between normal and cancer cells, and between human and rodent cells.

Cancerous Growth

The ability of cancer cells to grow excessively thus appears to be a consequence of several kinds of processes. One is the loss of a balance between cell proliferation and quiescence, normally regulated by the presence of nutrients and growth factors. If cells are usually quiescent, then their number cannot increase rapidly. If, on the other hand, cells are constantly stimulated to leave quiescence and enter the cycle, a more rapid increase in cell number will occur.

A second process is defective terminal differentiation. Cell number in many tissues is controlled and kept constant by a mechanism that causes some cells to depart permanently from the cycle when they undergo changes that favor their specialized function. Cancers of such tissues either ignore the signals to differentiate or are able to continue to divide after differentiation begins. Either aberration results in a net continuous increase in cell number.

The third process involved with excessive growth is the escape of cancer cells from a replicative senescence mechanism that limits the total number of divisions that normal cells can undergo, even if they do not remain quiescent or terminally differentiate. Escape from senescence allows cancer cells to grow indefinitely.

In chemotherapy and radiation therapy, drugs or X-rays are used to try to kill cancer cells at a faster rate than new cells can arise by division. Ideally, the cancer is eliminated. There are, however, many problems with chemotherapy, and eradication of cancer cells has proven very difficult. Nevertheless, for more advanced cases of cancer after metastases have occurred, localized treatments, such as surgery and radiation, no longer can cure, and chemotherapy remains the best treatment. Therapists need something that can specifically seek out and destroy all cancer cells. Immunotherapy is a hope for the future, as discussed elsewhere in this book.

Suggested Reading

General References

Cairns, J. *Cancer: Science and Society;* W. H. Freeman: San Francisco, 1978.

Hunter, T. "The Proteins of Oncogenes." *Scientific American* **1984,** *251 (August)*, 7-79.
Pardee, A. B.; Prem veer Reddy, G. *Cancer: Fundamental Ideas.* Carolina Biological Supply: Burlington, NC, 1986.
Sachs, L. "Growth, Differentiation and the Reversal of Malignancy." *Scientific American* **1986,** *254* (January), 30-37.

More Advanced Material

Alberts, B.; Bray, D.; Lewis, J.; Raff, M.; Roberts, K.; Watson, J. D. *Molecular Biology of the Cell;* Garland: New York, 1983; Chapter 11.
Baserga, R. *The Biology of Cell Reproduction;* Harvard University Press: Cambridge, 1985.
Darnell, J.; Lodish, H.; Baltimore, D. *Molecular Cell Biology;* W. H. Freeman: New York, 1986; Chapter 23.
Pardee, A. B. "Principles of Cancer Biology: Biochemistry and Cell Biology." In *Principles and Practice of Oncology*; DeVita, V. T., Jr.; Hellman, S.; Rosenberg, S. A., Eds; Lippincott: Philadelphia, 1985b; pp 3–22.
Pardee, A. B. *J. Cell Physiol.* **1985,** *Supplement 5*, 107–110.
Pardee, A. B.; Schneider, D. S. *Chemistry* **1977,** *50 (Jan./Feb.)*, 25.
Pitot, H. D. *Fundamentals of Oncology;* Marcel Dekker: New York, 1978.
Pratt, W. B.; Ruddon, R. W. *The Anticancer Drugs;* Oxford Press: Fairlawn, NJ, 1979.

CHAPTER 3 Cancer and the Cytoskeleton

Elaine Fuchs

The cytoplasm of a human cell is not simply a jellylike mass. It consists of an intricate array of structural elements, including a network of fibers known as the cytoskeleton. These fibers determine the shape or framework of the cell and, more importantly, may provide a network of communication between the cell surface, the cytoplasm, and the nucleus. Many outside signals, including growth factors and other environmental stimuli, appear to be transmitted from the plasma membrane through the cytoskeleton to the chromosomal DNA inside the nucleus. During neoplastic transformation, some of the interactions between the membrane and the cytoskeleton break down, contributing to miscommunication and uncontrolled growth.

Components of the Cytoskeleton

The cytoskeleton of all human cells is composed of three distinct fibrous networks: microfilaments (6 nm in diameter), microtubules (23 nm in diameter) and intermediate filaments (8–10 nm in diameter). All three structures are assembled from proteins that are encoded by families of about 10–50 genes.

1420–4/88/0037$06.00/0

Actins, the proteins that make up the microfilaments, are among the most highly conserved proteins in the evolutionary ladder. Almost 85% of the protein sequences of human and yeast actins are identical. Tubulins, the proteins that assemble into microtubules, are also highly conserved, with about 30% sequence identity between human and yeast tubulins.

In contrast, the intermediate filament (IF) proteins seem to be a highly diverse family; so far, no one has been able to establish a relationship between yeast and human IF proteins. Moreover, even human intermediate filament proteins vary dramatically from tissue to tissue. On the basis of their location in the body, IF proteins can be subdivided into five distinct classes: keratins exist in epithelial tissues, neurofilament proteins are found in cells of neural origin, glial filament protein is seen in glial cells and astrocytes, desmin seems to be exclusive to muscle cells, and vimentin is located primarily in fibroblasts. Thus, the intermediate filaments seem to be uniquely tailored to suit the specialized needs of each cell.

Microfilaments. Microfilaments are especially abundant in muscle cells, where they play a major role in the shape and contraction of muscle fibers. The discovery of actin in nonmuscle cells led to speculation that microfilaments might be involved in the shape and contractile movement of all cells. In the mid-1970s Elias Lazarides, who was then a graduate student under the guidance of James Watson at Harvard University, isolated the first antibodies against actin. With a fluorescent tag to the antibodies, he used a technique known as indirect immunofluorescence to visualize the internal network of microfilaments in a fibroblast cell. An example of this technique is shown in Figure 1a. The actin appears both as a filamentous net directly beneath the plasma membrane and as criss-crossed bundles of intracellular cables, called stress fibers. Other musclelike proteins (including myosin, tropomyosin and α-actinin) are often found with stress fibers. This suggests that these microfil-

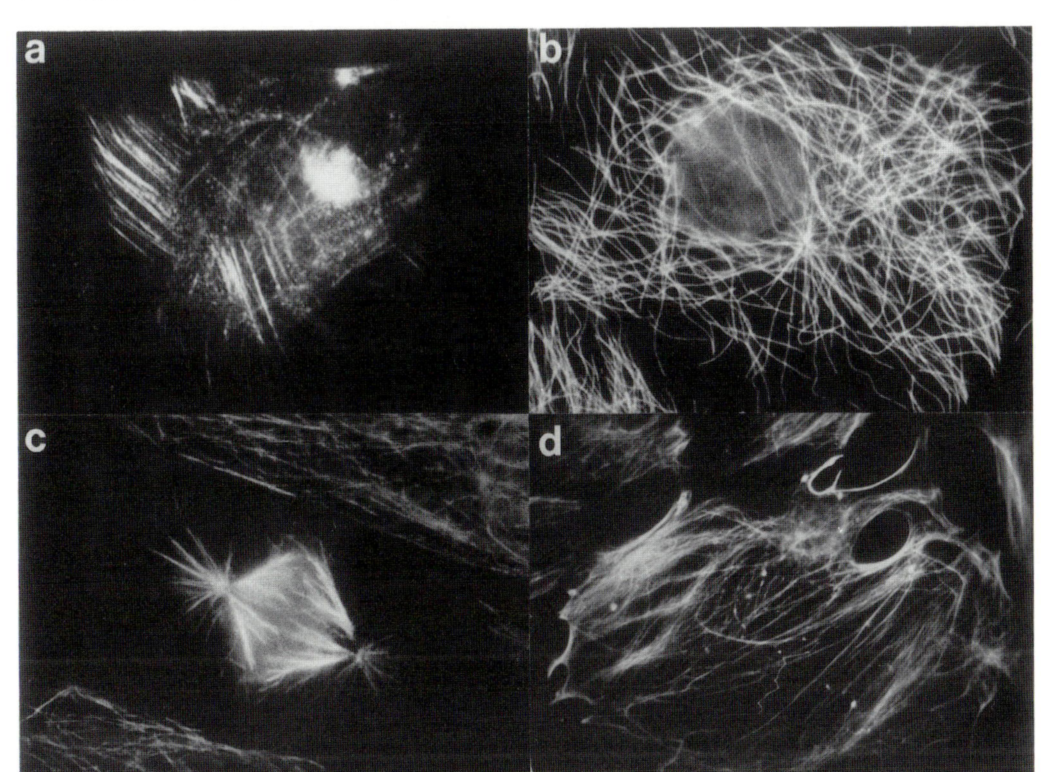

Figure 1. Major components of the cytoskeleton visualized by immunofluorescence. a: Actin stress fibers in a cultured rat fibroblast cell (courtesy, R. Kopan). b: Cytoplasmic microtubules in a cultured pig epithelial cell at interphase (courtesy, D. Vandre and G. Borisy). c: Spindle microtubules in a mitotic pig epithelial cell (courtesy, D. Vandre and G. Borisy). d: Vimentin intermediate filaments in a cultured rat fibroblast cell (courtesy, R. Kopan).

ament bundles may be the force underlying the mechanical movement of human cells (Lazarides and Revel, 1979). These fibers, anchored to the plasma membrane, could provide the support necessary for push and pull movements when the cell is adhered to a solid substrate.

Microtubules. Stress fibers play a major role in cell movement on solid surfaces, but microtubules seem to be more important in swimming movements (such as the beating action of eukaryotic cilia and flagella) and intracellular movements (including the mobility of cell surface receptors and the movement of pigment, secretory granules, and other organelles). (For a review, see Dustin, 1978.) Microtubules also contribute heavily to the structural framework of the cell, although their network is distinct from that of the actin filaments (Figure 1b). Just before mitosis, a major reorganization of the microtubules takes place. These structures temporarily direct their energy toward separating the duplicated daughter chromosomes into two identical nuclei (Figure 1c).

A number of proteins interact with microtubules (reviewed in Dustin, 1978). These include a family of high molecular weight (150–300 kd) microtubule-associated proteins (MAPs), some of which seem to be important in connections between microtubules and the other elements of the cytoskeleton. In neural cells, MAPs form protein crossbridges between the microtubules and the neurofilaments. In fibroblasts, similar interactions may take place between the microtubules and the vimentin filaments. Although crossbridges are not readily observed in fibroblasts, an interaction can be detected with a microtubule-destabilizing drug called colcemid. Treatment of fibroblasts with colcemid causes the vimentin network inside the cell to collapse and form a ring of IFs around the nucleus.

Intermediate Filaments. Microfilaments and microtubules have certain vital roles necessary for a cell's existence. In contrast, intermediate filaments seem to be essential only in multicellular organisms, where harmonious interactions between cells become important (Figure 1d). In fact, a few cell lines in tissue culture have been shown to lack intermediate filaments altogether. This fact confirms that, out of the context of the organism, cells do not need these structures for survival.

For cells that do have an intermediate filament network, it seems to be especially sensitive to environmental changes. Three different research teams, headed by James Rheinwald at Harvard Medical School, Avri Ben Ze'ev at the Weizmann Institute in Israel, and Elaine Fuchs at the University of Chicago, have contributed to the realization that some cultured cells can choose the type of intermediate filaments they will make, depending upon the nutrients provided to them and the degree of cell–cell interactions. Thus, in the presence of epidermal growth factor or when sparsely populated, cultured simple epithelial cells assume a fibrous structure and synthesize vimentin (Connell and Rheinwald,

1983; Ben-Ze'ev, 1984). In the presence of vitamin A and when allowed to come into close contact with each other, they become rounder and synthesize keratin (Kim et al., 1987). These findings suggest a possible role for intermediate filaments in achieving intercellular communication and organization within different tissues.

A few cell types may synthesize more than one kind of intermediate filament network, but most cells produce only a single type of IF protein. Even so, subtle adjustments to the intermediate filament network still occur in reponse to natural changes in the cell's environment. These changes seem particularly varied in the keratin-expressing epithelial tissues. Of all the intermediate filament classes, the keratins are by far the most diverse group. Humans produce more than 20 different keratin proteins, which vary widely in size (40–70 kd). Typically, only two to six keratins exist in an epithelial cell at any one time. However, as an epithelial tissue changes during development or differentiation, it synthesizes different sets of keratins. These changes might give rise to 8–10-nm filaments with different properties, such as solubility, tensile strength, and flexibility. Because the most variable portions of the keratin sequences protrude along the surface of the filament, filaments composed of different keratins may have different interactions with other proteins and other cytoskeletal elements inside the cell. In this way, the architecture of the cytoskeleton could be modified in accordance with fluctuations in cell–cell interactions within a tissue.

Keratin filaments serve an important protective function in the epidermis, and hence are especially abundant there. In the late 1970s at the Massachusetts Institute of Technology, Howard Green and I began to investigate the keratin network in epidermal cells at different stages of differentiation. In the innermost layer of the epidermis, a set of keratin proteins is formed and produces a dispersed cytoskeleton (Figure 2). An epidermal cell that migrates toward the skin surface ceases to divide and begins to make a new, somewhat larger, set of keratins. Reducing the level of vitamin A in the culture medium or exposing the epidermal cells to air and

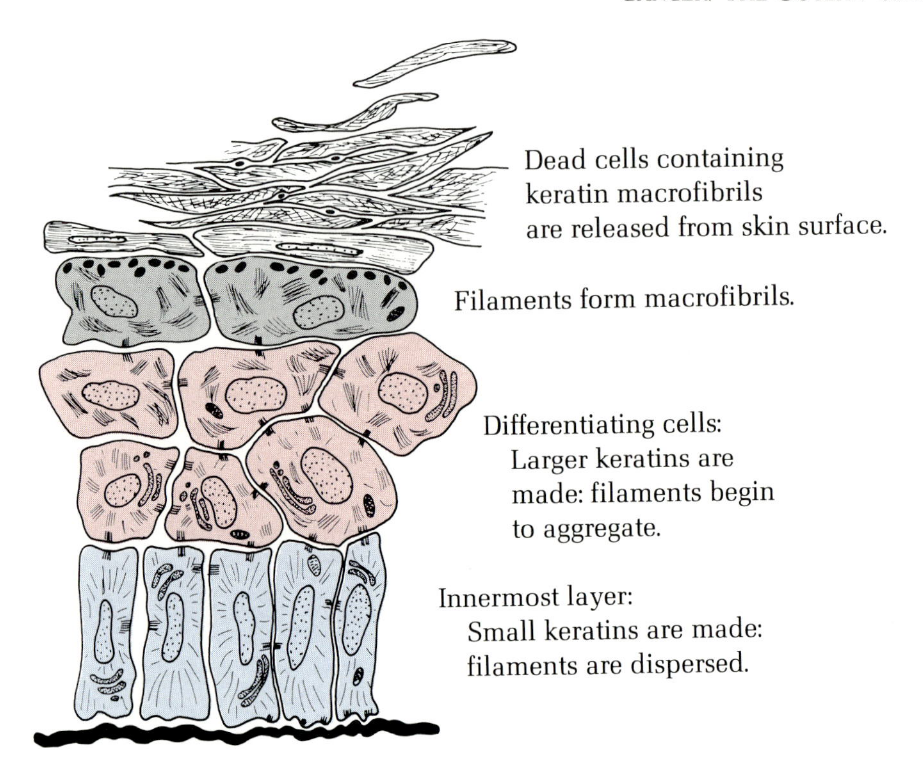

Figure 2. Terminal differentiation in the epidermis. As an epidermal cell leaves the innermost layer and begins to migrate outward toward the skin surface, it undergoes a variety of changes in its structure and its biochemistry.

feeding them from beneath will trigger both cell stratification and the production of large keratins in vitro.

Recently, Uli Aebi and his coworkers at the Johns Hopkins University demonstrated that these large keratins form aggregating filaments that could interfere with cell division, thereby contributing to differentiation. As the nondividing cell continues on its path toward the skin surface, the keratin filaments are bundled into structures called macrofibrils. By the time the epidermal cell is released from the skin, it is merely a dead sac densely packed with a tough, resilient cytoskeleton. Thus, keratin synthesis seems to adjust according to the growth rate, intercellular interactions and structural requirements of the epidermal cell.

Pathway of Communication

Between the Cell Exterior and the Cytoskeleton. Although the three networks of cytoskeletal filaments probably act together to bridge the communication gap, each has its own special function. The actin stress fibers receive signals from the exterior of the cell through localized points in the membrane, known as adhesion plaques. These protein-rich plaques correspond to the places on the surface of a cell that come into contact with a solid substrate (Figure 3). Some of the proteins in the adhesion plaque span the membrane, allowing communication between the inside and the outside of the cell. One of these proteins, vinculin, interacts with another protein, talin, to link the ends of the actin microfilaments to the inner side of the adhesion plaque (see Geiger et al., 1980; Pasquale et al., 1986).

The cell secretes a group of large sticky proteins, collectively known as the extracellular matrix (ECM), that help to anchor it to a solid surface. Fibronectin, one of the components of the ECM,

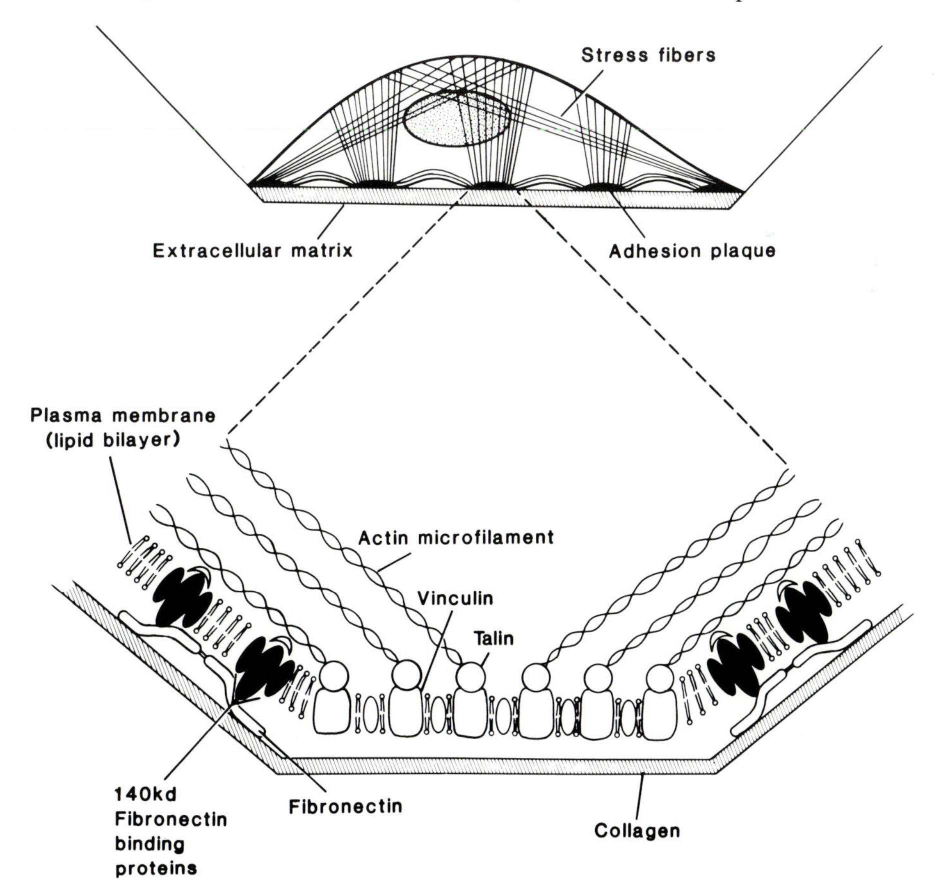

Figure 3. Components enabling cultured cells to adhere and stretch out on a solid surface. Top: Adhesion plaques are the points in the plasma membrane where a cell comes into contact with a solid surface. Actin stress fibers seem to emanate from adhesion plaques. Bottom: Expanded view of an adhesion plaque. Vinculin and talin are proteins that seem to play a role in anchoring the actin filaments to the membrane. Nearby, other membrane protein complexes anchor the fibronectin and the other components of the extracellular matrix to the cell surface.

binds the outer surface of the cell to a complex (140 kd) of three transmembrane proteins (Figure 3, reviewed in Hynes, 1986). In some cells, the position of the fibronectin on the outside of the cell is very similar to that of the actin stress fibers on the inside (Figure 4). Indeed, there must be either a direct or an indirect interaction between the membrane proteins that bind fibronectin and the ones that bind the actin stress fibers. This interaction can be shown by perturbing the actin filament network with a drug called cytochalasin B. Disruption of the microfilaments causes the release of fibronectin from the surface of the cell (Hynes, 1986).

Although the interactions between the plasma membrane and the microfilaments seem to be important for cell movement and attachment to a solid surface, interactions between the plasma membrane and intermediate filaments may play a role in intercellular communication. This possibility seems to be especially true for epithelial cells, where a network of keratin filaments extends to specific membrane sites, called desmosomes. Desmosomes represent zones of contact between adjacent epithelial cells. The desmosomal proteins in the membrane of one cell interact directly with those in the membrane of an adjacent cell. In this way, a direct line of communication can be established throughout all of the epithelial cells of a tissue (Figure 5; reviewed in Matoltsy, 1986).

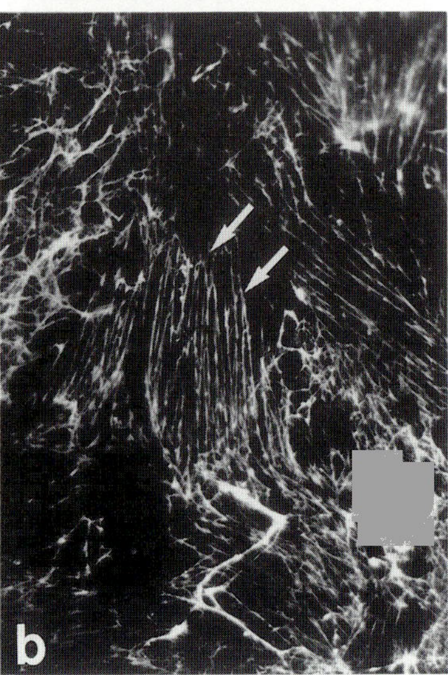

Figure 4. Fibronectin and actin frequently are arranged in tandem. Antibodies to actin (a) and fibronectin (b) reveal that the fibronectin fibrils on the outside of the cell are aligned in concert with the actin stress fibers on the inside of the cell (Reprinted with permission from Hynes and Destree. Copyright 1978 Cell Press).

Between the Cytoskeleton and the Nucleus. The nuclear envelope separates the cytoplasm from the chromosomal DNA, and thus channels all transport across the nuclear membrane through specialized structures called nuclear pores. The major structural component of the nuclear envelope is the nuclear lamina, a meshwork of 10-nm filaments on the inner side of the nuclear envelope (Figure 6; reviewed in Krohne and Benavente, 1986). The nuclear lamina comprises proteins called lamins, which only recently have been identified as members of the intermediate filament family. The nuclear lamina may associate directly with the chromosomes when genes are actively being transcribed.

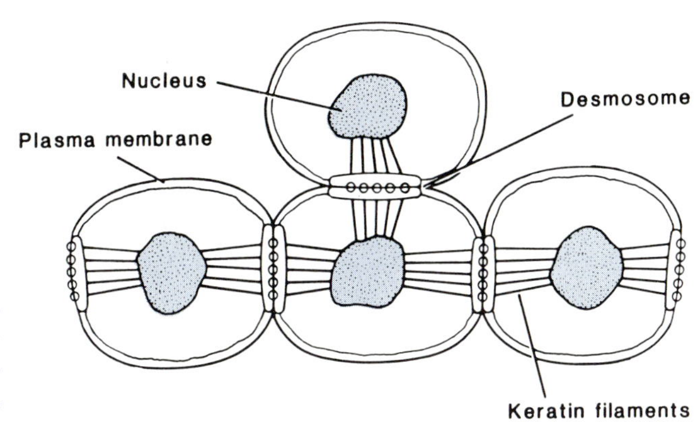

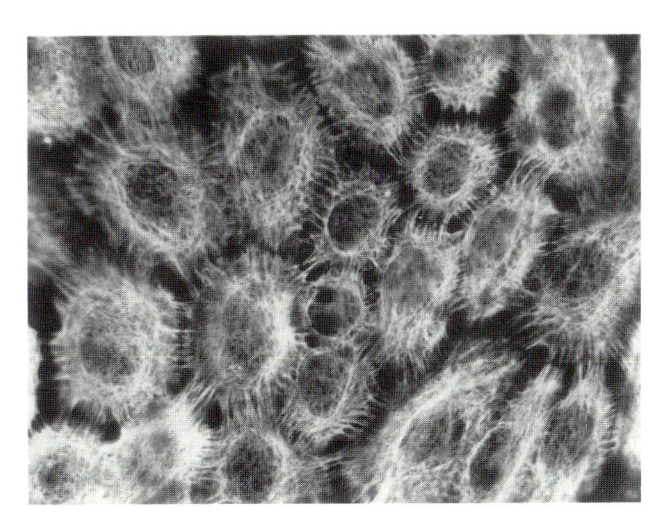

Figure 5. Keratin filaments are anchored to membrane sites known as desmosomes. Top: Schematic. Desmosomes are protein complexes which span the plasma membranes of adjacent epithelial cells. On the inside of the cell, the desmosomes seem to attach to the keratin filaments, and thereby establish a link between interacting cells and the cytoskeleton. Bottom: Immunofluorescence photograph. Cultured epidermal cells stained with an anti-keratin antibody (courtesy, R. Kopan).

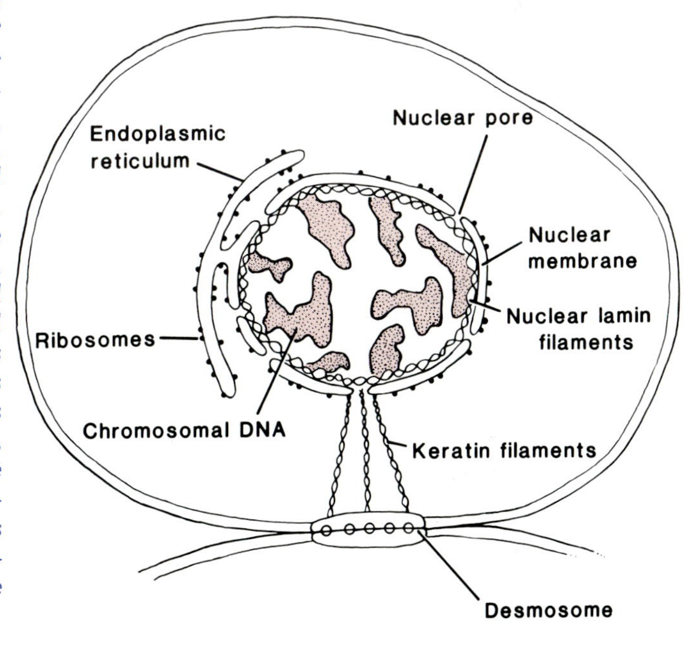

Figure 6. Does the cytoskeleton interact with the chromosomes through the nuclear envelope? On the inner side of the nuclear membrane is a matrix of intermediate filamentlike fibrils, made of proteins called lamins. This matrix is closely associated with actively transcribed genes. Outside the nuclear envelope are the intermediate filaments, which are closely associated with the nucleus. In epithelial cells, if the keratin filaments interact with the lamin filaments either through the nuclear pores or via the nuclear membrane, this interaction could provide the final link of a putative pathway between the chromosomes inside the nucleus and the desmosomes that interconnect the epithelial cells within a tissue.

In addition to interactions with chromosomes, the lamina may also form a scaffold on which all of the nuclear pores are attached. Outside the nucleus, the cytoplasmic intermediate filaments are frequently close to the nuclear envelope. The cytoplasmic intermediate filaments might possibly interact with the lamins, either directly via the nuclear pores or indirectly via a component of the nuclear envelope. Because interactions have been identified between the cytoplasmic intermediate filaments and the microfilaments and microtubules, the cytoskeleton may provide a series of interconnected lines of communication from external stimuli to the cell nucleus.

Just prior to mitosis, the nuclear membrane disintegrates. This allows the cytoplasmic microtubules to gain access to the chromosomes, so that they can form the spindle network necessary for separation of the daughter chromosomes. Thus, even when the nuclear lamina temporarily disappears, the hereditary information in the cell is always closely associated with components of the cytoskeleton.

Cell Shape, Growth Control, and Cancer

It has been known since the early 1970s that cells grow better when they attach and are able to "stretch out" on a solid support. This phenomenon is known as anchorage dependence. Normal cells that are grown to high density in tissue culture do not have room to stretch out, and they begin to retract their proliferation processes and round up. A classical study by Judah Folkman and Anne Moscona at the Harvard Medical School suggested that normal tissue culture cells may stop proliferating when they round up because they can no longer respond to extracellular stimuli such as growth factors (Folkman and Moscona, 1978). A later study conducted at The Massachusetts Institute of Technology demonstrated that changes in the shape of the cell induce alterations in the cytoskeleton. In turn, these alterations seem to disrupt signals sent from the membrane to the nucleus (Ben-Ze'ev et al., 1980). Indeed, when the growth rate of a cell changes, the cytoskeleton frequently undergoes gross reorganization, accompanied by an alteration in the expression of the corresponding cytoskeletal proteins.

During neoplastic transformation, cells lose their anchorage dependence and begin to grow in suspension. They become less responsive to outside growth factors and to changes in cell shape, and their growth becomes uncontrolled. The cells do not stop proliferating at high cell density, and they gradually assume an abnormal structure, accompanied by alterations in the cytoskeleton (Wittelsberger et al., 1981). Communication between the cell exterior and the nucleus seems to be either lost or misfiring.

Microfilament Network. In 1975–6, experiments conducted by Klaus Weber's group at the Max Planck Institute in Gottingen, Federal Republic of Germany, and S. Jonathan Singer's group at The University of California in San Diego demonstrated an absence of stress fibers in the cytoplasm of some cancerous cells grown in tissue culture. The disappearance of stress fibers can be observed by transforming normal tissue culture cells in vitro. One of the most widely used models of neoplastic transformation has been the infection of chick embryo fibroblast cells with a chicken tumor-promoting virus, known as Rous sarcoma virus (RSV). In RSV-transformed cells, the overall level of actin does not appear to decrease. Rather, the actin is redistributed to points located beneath the plasma membrane (Figure 7; compare b and d; also see Nigg et al., 1986).

The disappearance of stress fibers seems to accompany a loss of cell adhesiveness to the substrate. The poor adherence of transformed cells seems to stem from inability to produce fibronectin and organize it into extracellular matrix. If the substrate is artificially coated with fibronectin, even cancer cells will adhere to it, spread out (Hynes, 1986), and reestablish their internal network of stress fibers.

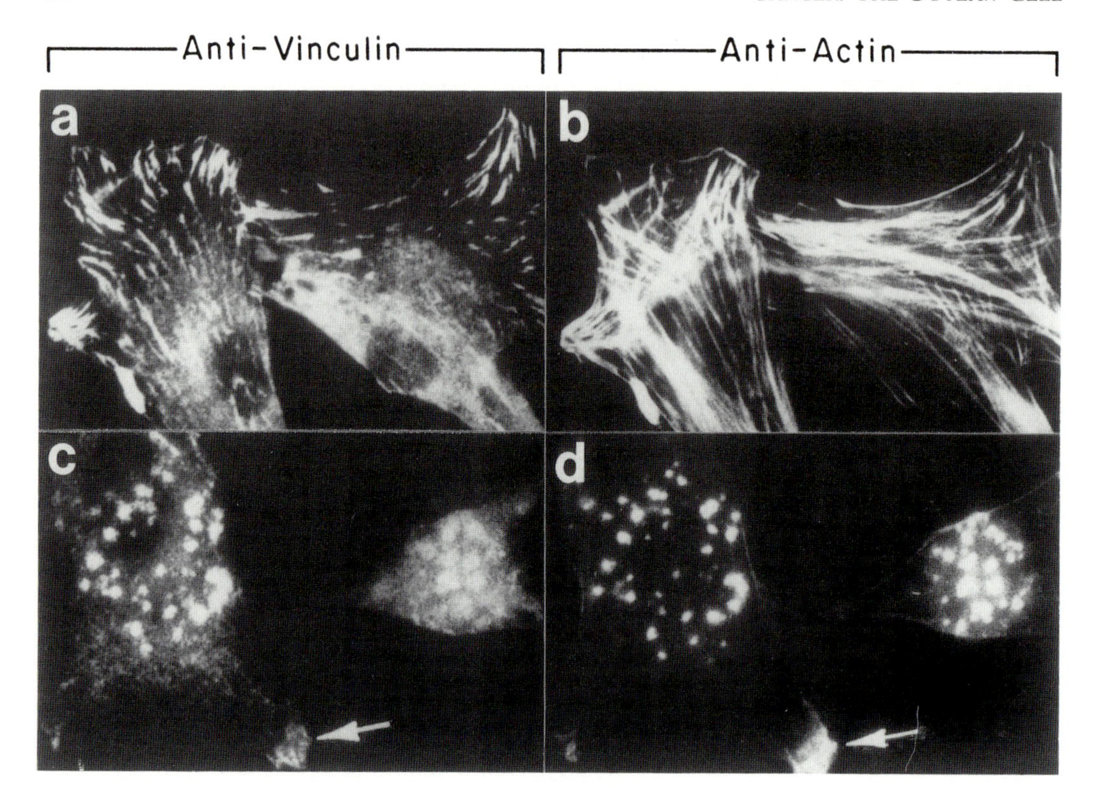

It is not yet clear if the reorganization of actin and the loss of cell-surface fibronectin seen in tissue culture cells plays any direct role in the invasiveness of tumor cells. Transformed cells frequently have reduced extracellular fibronectin, deteriorated microfilament bundles, and other prominent changes in the communication pathway between the cell surface and the nucleus. Most of the focal adhesion sites, which form the link between fibronectin and stress fibers, disappear. Vinculin and talin reorganize into smaller contact sites, which may provide considerably weaker anchorage than true focal adhesion sites (Figure 7; compare a and c; Pasquale et al., 1986; Nigg et al., 1986).

Figure 7. The organization of actin and vinculin is frequently disrupted when fibroblasts are transformed in culture. In normal chick embryo fibroblasts, vinculin (a) is located with actin (b) on the inner surface of the cell membrane. When these cells are transformed with RSV, the vinculin (c) and actin (d) maintain their association, but are redistributed in rosette-like structures. Actin and vinculin reorganize in a variety of transformed cells in culture, but their redistribution is not always into rosette patterns (Reprinted with permission from Nigg et al. Copyright 1986 Academic Press).

Two major groups, one headed by Ray Erikson, now at Harvard University, and the other by J. Michael Bishop at the University of California, San Francisco, Medical School, have shown that a single gene carried by the RSV virus is responsible for promoting tumors. This marked the first discovery of a cancer-causing gene, or oncogene. The protein encoded by this gene, referred to as $p60^{src}$, is an enzyme kinase capable of phosphorylating tyrosine residues. Although most of the phosphorylation in normal cells occurs at serine or threonine residues of proteins, substantial levels of phosphorylated tyrosine residues are found in some tumor cell proteins (for review, see Cooper and Hunter, 1983). The tyrosine phosphorylation of some of these proteins may lead to malignancy.

Tony Hunter and coworkers at the Salk Institute showed that vinculin is one of the few proteins that normally contains phosphotyrosine. During viral transformation, however, the level of tyrosine phosphorylation on vinculin increases almost 20-fold (Sefton et al., 1981). The use of fluorescent antibodies to phosphotyrosine confirmed that adhesion plaques are a major early target of the $p60^{src}$ oncogene protein. Shortly after phosphorylation, the adhesion plaques seem to disintegrate.

Vinculin is not the only adhesion plaque protein that becomes phosphorylated with tyrosine in tumor cells. S. Jonathan Singer's laboratory has recently shown that while talin is not tyrosine-phosphorylated in normal cells, three times more talin than vinculin is phosphorylated in RSV-transformed cells. Another target for tyrosine phosphorylation is p36, a protein whose function is still unknown. This protein is present only in some cells of the body, where it seems to be associated with both the inner membrane and the actin cytoskeleton (Cooper and Hunter, 1983). Finally, most tyrosine kinases, including $p60^{src}$, can phosphorylate themselves. They concentrate in the weak contact points between the transformed cell and its substrate.

Because RSV-transformed cells show all of the cytoskeletal disruptions characteristic of tumor cells, it is tempting to speculate that phosphorylation at adhesion plaques may lead directly to the transformed state, perhaps through blockage of fibronectin- or actin-binding sites. Alternatively, $p60^{src}$ itself may be involved in this blockage. Whatever the mechanism, it is unlikely that the mere increase in tyrosine kinase activity in a cell is crucial. Mutants of $p60^{src}$ retain their full tyrosine

kinase activity but cannot bind to cell membranes or transform cells (Kamps et al., 1986).
(Kamps et al., 1986).

Although transformation through expression of membrane-bound tyrosine kinases may be one way of triggering cytoskeletal disruptions that lead to loss of growth control, it certainly does not seem to be the only way. Recently, several interesting reports have suggested that changes in actin itself can elicit these responses. For instance, an aberrant β-actin was detected in human cells transformed by chemical mutagens (Leavitt et al., 1982). The mutant actin may disrupt the polymerization or organization of microfilaments and thereby cause tranformation. In another case, the oncogene of a feline transforming virus, fgr, was shown to encode a hybrid protein consisting of a 128-amino-acid-residue segment of actin fused to a segment of a protein bearing a strong resemblance to the active site portion of a tyrosine kinase (Naharro et al., 1984). The hybrid actin may either interfere directly with microfilament organization or guide its tyrosine phosphorylating activity to the focal adhesion plaques. Finally, the expression of actin genes is one of the fastest cellular responses to extracellular growth factors. Therefore, it seems reasonable to expect that alterations not only in the actin protein, but perhaps also in its synthesis, might lead to malignant transformation.

Intermediate Filament Network. Substantial evidence indicates that the intermediate filament network is altered in malignant cells. S. Jonathan Singer's group has reported that the association of microtubules and intermediate filaments is disrupted in RSV-transformed fibroblasts, and this distruption causes the intermediate filaments to retract to a perinuclear region (Ball and Singer, 1981). In a separate study, Edward Fey and Sheldon Penman discovered a progressive change in the organization of the nuclear matrix and the cytoplasmic intermediate filaments in epithelial cells following treatment with a tumor-promoting agent known as TPA (Fey and Penman, 1984). With longer treatment, desmosomal complexes also begin to dissociate.

In addition to changes in IF distribution, the formation of IF proteins can also be abnormal. Ying-Ji Wu and James Rheinwald were the first to discover a marked difference in keratin synthesis between normal and malignant epidermal cells (Wu and Rheinwald, 1981). Although cells continue to divide in basal cell carcinomas of the skin, they never differentiate or express the large keratins associated with this process. In squamous cell carcinomas of the skin, the cells undergo only partial terminal differentiation, and hence produce only low levels of large keratins. In squamous cell tumors, an entirely new group of keratins appears.

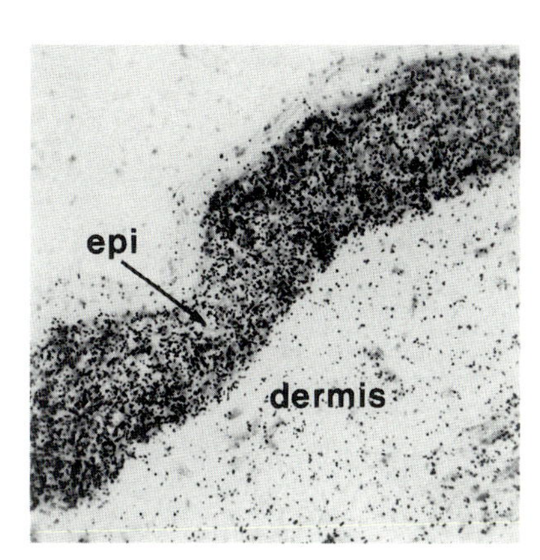

Figure 8. Messenger RNAs encoding a hyperproliferation-associated keratin are present in human skin, even though the protein is made only during wound-healing or in squamous cell carcinomas. A radiolabeled probe was used to detect an mRNA encoding one of the keratins that is expressed transiently in wound-healing (constitutively in skin cancers). The black grains indicate the presence of mRNAs for this keratin (Reprinted with permission from Tyner and Fuchs. Copyright 1986 Rockefeller University Press).

Henry Sun and his coworkers at The New York University School of Medicine (Weiss et al., 1985) and Birgitte Lane at the Imperial Cancer Research Fund in London (personal communication) have shown that the new keratins regularly synthesized in squamous cell tumors appear transiently in normal skin during wound-healing and hyperproliferation. Using a technique known as in situ hybridization, Angela Tyner in my laboratory has shown that the RNAs encoding the keratins associated with hyperproliferation are always present in skin, even though the proteins are only transiently made (Figure 8; Tyner and Fuchs, 1986). Thus, the normal epidermal cell may have devised a method to produce these proteins rapidly when needed, from preexisting RNAs. We do not yet know why a cell might need these keratins during hyperproliferation. The keratin filaments assembled from this set of proteins may enable cells to either undergo more cell divisions prior to terminal differentiation or divide at a faster rate.

Because the pattern of keratins seems to be closely tuned to the particular differentiated state of a cell, and because keratin expression seems to be highly dependent upon intercellular interactions, it is not surprising that the expression of keratins should be altered during neoplastic transformation. This observation is true for many types of epithelial cells that have been transformed either naturally, with oncogenic viruses, or with tumor-promoting agents (for a review, see Moll et al., 1983). The mechanism of altering keratin production in malignancies is not yet known. In some cases, it seems that the cancer cell has acquired increased sensitivity to extracellular regulators of differentiation, such as vitamin A (Kim et al., 1984). In other cases, the cell may lose its sensitivity to growth regulators, such as epidermal growth factor. The question of whether the alteration in keratin synthesis follows or precedes other changes in the cytoskeleton awaits further examination.

Cancer Diagnosis and Treatment

The importance of the cytoskeleton in cell growth and nuclear division has led to the development of a major class of drugs for cancer treatment. Because cancer cells grow uncontrollably, inhibitors of cell division are powerful tools in chemotherapy. Drugs that interfere with microtubule formation show marked antimitotic activity. Colchicine, a plant alkaloid, is one of the best studied of the antimitotic drugs. It disrupts microtubule polymerization and therefore interferes with spindle formation. Vinblastine, vincristine, and vindesine also interfere with microtubule assembly, but they do so by crystallizing the tubulin protein inside the cell. These drugs have a broad spectrum of antitumor activity and have been widely used clinically for people with advanced or metastatic carcinomas. Taxol and nocodazole, microtubule inhibitors that have proven to be very useful in animal studies, are now beginning to be used in clinical trials.

An increased understanding of the cytoskeleton and its components has led to the development of anticancer drugs and has also provided new and improved methods for cancer diagnosis. In this regard, antibodies specific for different intermediate filament proteins have been particularly useful to pathologists in confirming their diagnoses. Mary Osborn at the Max Planck Institute in Gottingen, Federal Republic of Germany, and Roland Moll and Werner Franke at the German Cancer Research Center in Heidelberg, Federal Republic of Germany,

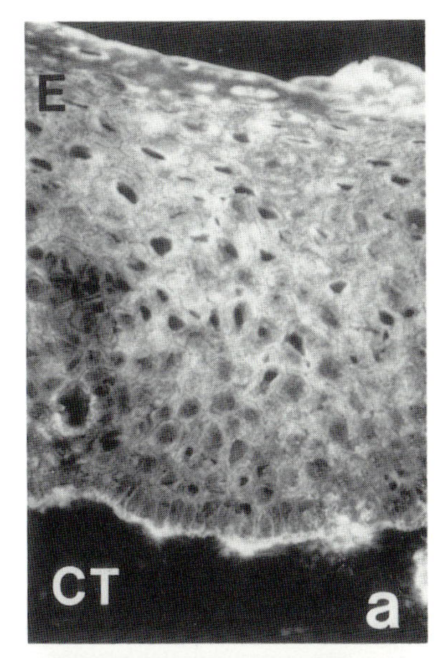

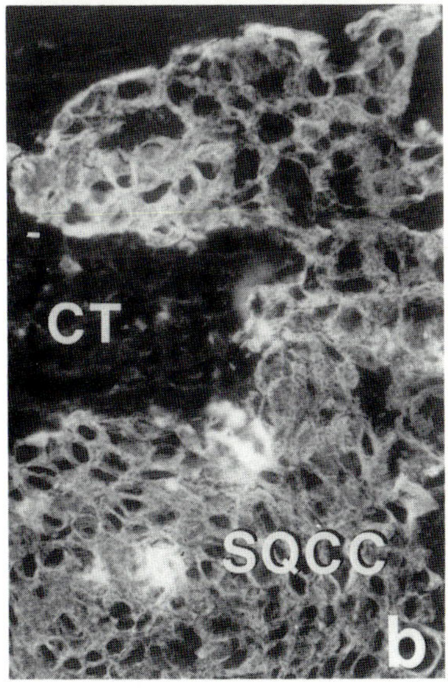

Figure 9. Antibodies to keratin filaments identify cells of epithelial origin in both normal and malignant tissues. An antibody against keratin was used to detect the epithelial cells in normal human esophageal tissue (a) and in squamous cell carcinoma (SQCC) of the esophagus (b). The connective tissue (CT) containing mostly fibroblasts does not recognize the antibody. (Reprinted with permission from Grace. Copyright 1985 Cancer Research)

realized in the course of their studies that neoplastic cells continue to express the same type of intermediate filament as their normal counterpart. This discovery led to the development of antibodies to each of the five intermediate filament classes. By determining which of these fluorescently tagged antibodies reacts with the tumor, a pathologist can readily identify or confirm the cell type from which the tumor originated (Figure 9; see also Moll et al., 1983, and Osborn and Weber, 1983). This identification can be especially helpful when tumors are difficult to diagnose or have metastasized.

For carcinomas, that is, cancers originating from epithelial cells, the use of antibodies to keratins may eventually provide the pathologist with an even more refined set of tools. Because various subsets of the more than 20 individual keratins are expressed in different epithelial cells, a set of monoclonal antibodies, each specific for a single keratin, should help to determine the precise origin of the tumor cell (Huszar et al., 1986). Moreover, because many epithelial tumors produce a well-defined but altered set of keratins upon malignant transformation, some monoclonal antibodies may eventually be useful in distinguishing between malignant and nonmalignant tissue. As researchers conduct more detailed studies, new and additional diagnostic tools should continue to surface.

Acknowledgments

I especially thank Richard Hynes (Massachusetts Institute of Technology) for providing double immunofluorescence photographs of cells stained with anti-actin and anti-fibronectin; S. Jonathan Singer (University of California, San Diego) for providing double immunofluorescence photographs of normal and transformed cells stained with anti-actin and anti-vinculin; Dale Vandre and Gary Borisy (University of Wisconsin) for providing immunofluorescence photographs of interphase and mitotic cells stained with anti-tubulin; Raphael Kopan (my laboratory) for providing immunofluorescence photographs of rat dermal papilla fibroblasts stained with anti-actin and anti-vimentin and cultured rat epidermal cells, stained with anti-keratin; and Philip Galiga and David Rosenzweiz (Art Department, University of Chicago) for their expert assistance in preparation of the figures and drawings.

Suggested Reading

Ball, E. H.; Singer, S. J. *Proc. Natl. Acad. Sci. USA* **1981,** 6986–6990.

Ben-Ze'ev, A. *J. Cell Biol.* **1984,** *99,* 1424–1433.

Ben-Ze'ev, A.; Farmer, S. R.; Penman, S. *Cell* **1980,** *21,* 365–372.

Connell, N. D.; Rheinwald, J. G. *Cell* **1983,** *34,* 245–253.

Cooper, J. A.; Hunter, T. *Curr. Top. Microbiol. Immunol* **1983,** *107,* 125–161.

Dustin, P. *Microtubules;* Berlin: Springer–Verlag, 1978.

Fey, E. G.; Penman, S. *Proc. Natl. Acad. Sci. USA* **1984,** *81,* 4409–4413.

Folkman, J.; Moscona, A. *Nature (London)* **1978,** *273,* 345–349.

Geiger, B.; Tokuyasu, K. T.; Dutton, A. H.; Singer, S. J. *Proc. Natl Acad. Sci. USA* **1980,** *77,* 4127–4131.

Grace, M. P.; Kim, K. H.; True, L. D.; Fuchs, E. *Cancer Res.* **1985,** *45,* 841–846.

Huszar, M.; Gigi-Leitner, O.; Moll, R.; Franke, W. W.; Geiger, B. *Differentiation (Berlin)* **1986,** *31,* 141–153.

Hynes, R. O. *Sci. Am.* **1986,** *254,* 42–51.

Hynes, R. O.; Destree, A. T. *Cell* **1978,** *15,* 875–886.

Kamps, M. P.; Buss, J. E.; Sefton, B. M. *Cell* **1986,** *45,* 105–112.

Kim, K. H.; Schwartz, F.; Fuchs, E. *Proc. Natl. Acad. Sci. USA* **1984,** *81,* 4280–4284.

Kim, K. H.; Stellmach, V., Javors, J.; Fuchs, E. *J. Cell Biol.* **1987,** *105,* 3039–3052.

Krohne, G.; Benavente, R. *Exp. Cell Res.* **1986,** *162,* 1–10.

Lazarides, E.; Revel, J. P. *Sci. Am.* **1979,** *240,* 100–113.

Leavitt, J.; Bushar, G.; Kakunaga, H.; Hamada, H.; Hirakawa, T.; Goldman, D.; Merril, C. *Cell* **1982,** *28,* 259–268.

Matoltsy, A. G. In *Biology of the Integument Volume 2*; Springer-Verlag: Berlin, 1986; pp 255–361.

Moll, R.; Krepler, R.; Franke, W. W. *Differentiation (Berlin)* **1983,** *23,* 256–269.

Naharro, G.; Robbins, K. C.; Reddy, E. P. *Science (Washington, DC)* **1984,** *223,* 63–66.

Nigg, E. A.; Sefton, B. M.; Singer, S. J.; Vogt, P. K. *Virology* **1986,** *151,* 50–65.

Osborn, M.; Weber, K. *Lab. Invest.* **1983,** *48,* 372–394.

Pasquale, E. B.; Maher, P. A.; Singer, S. J. *Proc. Natl. Acad. Sci. USA* **1986,** *83,* 5507–5511.

Sefton, B. M.; Hunter, T.; Ball, E. H.; Singer, S. J. *Cell* **1981,** *24,* 165–174.

Tyner, A. L.; Fuchs, E. *J. Cell Biol.* **1986,** *103,* 1945–1955.

Weiss, R.A.; Eichner, R.; Sun, T.-T. *J. Cell Biol.* **1985,** *98,* 1397–1406.

Wittelsberger, S. C.; Kleene, K.; Penman, S. *Cell* **1981,** *24,* 859–866.

Wu, Y.-J.; Rheinwald, J. G. *Cell* **1981,** *25,* 627–635.

CHAPTER 4 Cancer, Viruses, and Oncogenes

Arnold J. Levine

Viruses are intracellular parasites whose submicroscopic size and relative genetic simplicity have attracted the interest and imagination of microbiologists for almost 100 years. They cannot be seen in the normal light microscope and pass through filters that retain most or all known bacteria.

In spite of their modest proportions, viruses are responsible for a wide variety of diseases, both ancient (polio, smallpox) and modern (AIDS). They continue to provide new and important challenges to the population and the scientist. Many of these viruses cause cancer in animals, and some viruses are closely associated with human cancers (Davis et al., 1980). In these cases, the tumor-causing virus becomes an excellent experimental tool that will lead to increased understanding of these uncontrollable diseases.

Tumor Virology

Today viruses are classified by their chemical composition and the way in which they reproduce, hide, or function in their life cycle. These agents have been divided into two "chemical kingdoms"

1420–4/88/0057$06.00/0

according to the chemical composition of their genetic information, deoxyribonucleic acid (DNA) or ribonucleic acid (RNA). Using these criteria, virologists have been able to classify thousands of agents in each kingdom into different virus families (Davis et al., 1980). The names of these families are presented in List I, along with some common examples of human pathogens.

Almost every DNA virus family mentioned in List I has a representative that can produce tumors in animals, that is, cause cancer. Most are tumor viruses under the appropriate and often specialized circumstances of an experiment. In contrast, among the thousands of RNA viruses in the families listed, only the retroviruses contain examples of tumor viruses. The retroviruses, which convert their RNA genome into a DNA genetic intermediate step during their life cycle, are intermediate between DNA and RNA viruses. They bridge the chemical kingdoms.

List I. Classification of Viruses

DNA Viruses	*RNA Viruses*
1. Parvoviruses: Adeno-associated virus	1. Picornaviruses: poliovirus
2. Papovaviruses:	2. Orthomyxoviruses: Influenza virus
a. Polyoma-SV40 group: SV40	3. Paramyxoviruses: Measles virus
b. Papillomaviruses: Human wort virus	4. Coronaviruses: Coronaviruses
3. Adenoviruses: Adenoviruses	5. Rhabdoviruses: Rabies virus
4. Herpesviruses: Herpes simplex I, II	6. Toga, Bunya, Arenaviruses: Yellow fever, Rift Valley fever, Lassa fever
5. Poxviruses: Smallpox	7. Retroviruses: Rous Sarcoma virus
6. Hepadnaviruses: Hepatitis B	

In DNA tumor viruses, R. E. Shope first recognized the viral origin of some tumors (Shope, 1933). He isolated an agent from benign cutaneous tumors or warts (called papillomas) of the wild rabbit. The virus passed through filters that normally retained bacteria and could reproducibly induce the same benign tumor at the site of injection in both wild and domestic rabbits. This viral agent also was able to produce a new malignant disease, a carcinoma, but only in domestic rabbits (a different species from the wild rabbit). In addition, Shope was unable to isolate the infectious virus from this carcinoma.

These early observations, that a virus isolated from one species of animal could produce malignant disease in a different species and that the inoculating infectious virus could no longer be isolated from the tumor tissue, were to form a reproducible and standard pattern in the field of DNA tumor virology. In almost all cases, only part of the genetic information carried by these DNA tumor agents functions in the transformation of a normal cell into a malignant cell. Thus a transformed cell may contain only a fragment of the entire viral genome, and infectious virus can no longer be rescued from such cells. These transforming viral genes (Table I) have, in many cases, been identified and are unique to the viral agent. Transforming genes and gene products from a wide variety of different viruses may have some functions in common, but they are chemically and genetically unrelated to each other (Varmus and Levine 1983; Vande Woude et al., 1984).

Table I. DNA Tumor Viruses: Viral Oncogenes

	Genome Size		
Viruses	*Kb*	*No. Genes*	*No. Viral Oncogenes*
Papovaviruses			
Polyoma–SV40	5.5	5–6	1–3 genes; T, m-T, t
Papilloma	8	10	2 genes; E5, E6
Adenoviruses	36	50	2–4 genes; E1A, E1B
Herpesviruses	110–180	80–120	EBV: 1–3 genes; LMP, EBNA-II
Poxviruses	220–240	100–200	unclear (EGF-like gene known)
Hepadnaviruses	3.2	4–6	unclear
RNA tumor viruses—oncogenes			
Retroviruses			
Slow neoplasia	5–10	3	none, promotor-insertion in the host cell chromosome adjacent to c-onc
Acute-transforming	4–10	3–5	1–2 from derived proto-oncogene of the host cell

Whatever the origins of these viral genes in the distant past, the transforming viral cancer-causing genes (oncogenes) are dissimilar to contemporary cellular genes and to the genetic information of the host cell. Each virus group has its own set of viral oncogenes. This point is important because the transforming genes of the acute RNA tumor viruses originate in the normal cellular genome of the host cell. A normal cellular gene, termed a proto-oncogene (it does not cause cancer in the normal host), is picked up by the retrovirus, incorporated into its viral chromosome, and frequently altered by mutation. It then becomes an oncogene, able to contribute to the cause of a cancer (Varmus and Levine, 1983; Vande Woude et al., 1984). The retrovirus supplies the oncogene's ability to spread from cell to cell and from host to other animals, to produce an "infectious cancer."

Thus, we can divide the viral agents by origin into two distinct groups that can produce tumors in animals. The retroviruses (RNA tumor viruses) intimately interact with—indeed, incorporate into their own genome—the proto-oncogenes of the host cell. When a proto-oncogene is activated or altered in some way, it becomes an oncogene and its product can cause a cancer. The DNA tumor viruses, on the other hand, encode in their own chromosomes the genetic information to alter cellular growth control. There is no chemical relationship between these DNA viral oncogenes and the cellular proto-oncogenes or RNA virus oncogenes.

We have identified the oncogenes and demonstrated that the viral and cellular oncogenes contribute significantly to the cause of virus-induced cancers. The next question concerns the function of the protein products of these oncogenes. How does altering their function or regulation cause cancer? How can we stop this altered function without preventing normal function? Even though the viral and cellular oncogenes are genetically and chemically unrelated in nucleotide sequence, they may share functions. Equally interesting, the proteins may interact with each other to effect a new function. This article will explore the mechanisms by which DNA and RNA tumor viruses act to produce cancer. Along the way we will get our first clues about the molecular basis of growth control and uncontrolled growth.

Retroviruses—The Cellular Origins of Oncogenes

Retroviruses carry and transmit cellular derived oncogenes. These viruses can incorporate cellular genetic information into a viral chromosome and move this information from cell to cell or host to host. Historically, this is how oncogenes were first recognized. It will be useful, therefore, to review how these viruses reproduce, incorporate oncogenes, and spread them to other cells.

Growth Cycle. Figure 1 illustrates the retrovirus replication cycle. Retroviruses are enveloped particles 90–100 nm in diameter, with prominent spikes composed of virus-encoded glycoproteins projecting from a lipid bilayer. These envelopes surround a ribonucleoprotein particle composed of two identical single-stranded RNA subunits (a diploid genome) 5–10 Kb in size. Reverse transcriptase is an enzyme closely associated with the ribonucleoprotein particle. It copies RNA into a faithful DNA duplex. Structural proteins package the RNA genome (*gag*) (Davis et al., 1980). These three viral proteins, the glycoprotein envelope (*env*), ribonucleoprotein core (*gag*) and reverse transcriptase (*pol*) compose the three essential genes and gene products of these viruses. They form a linear RNA chromosome surrounded on both sides by regulatory signals (called long terminal repeat signals or LTRs) designed to ensure the production of large levels of RNAs in the infected host cell. The LTR is composed of an "enhancer–promotor" signal (a sequence of nucleotides) recognized by the host cell enzymatic machinery to efficiently reproduce the virus. Retroviruses attach to their host cell using the surface glycoprotein (*env*), and the ribonucleoprotein core enters the cell via a virus envelope–cell membrane fusion. The reverse transcriptase makes a duplex DNA copy of the RNA genome. The copy enters the nucleus and integrates, apparently in a random fashion, into a chromosome of the host cell. The strong regulatory signals in the LTR of the viral chromosome are read, and high levels of viral *m*-RNA's are transcribed. Some of this RNA is used to translate proteins on the ribosomes, and more *gag*, *pol*, and *env* gene products are produced. The *gag*-*pol*–RNA complexes move to the plasma membrane of the cell, which already contains *env*-glycoprotein inserted into it. A particle forms and buds out into the fluids surrounding the cell. This new particle is now infectious (Varmus and Levine, 1983).

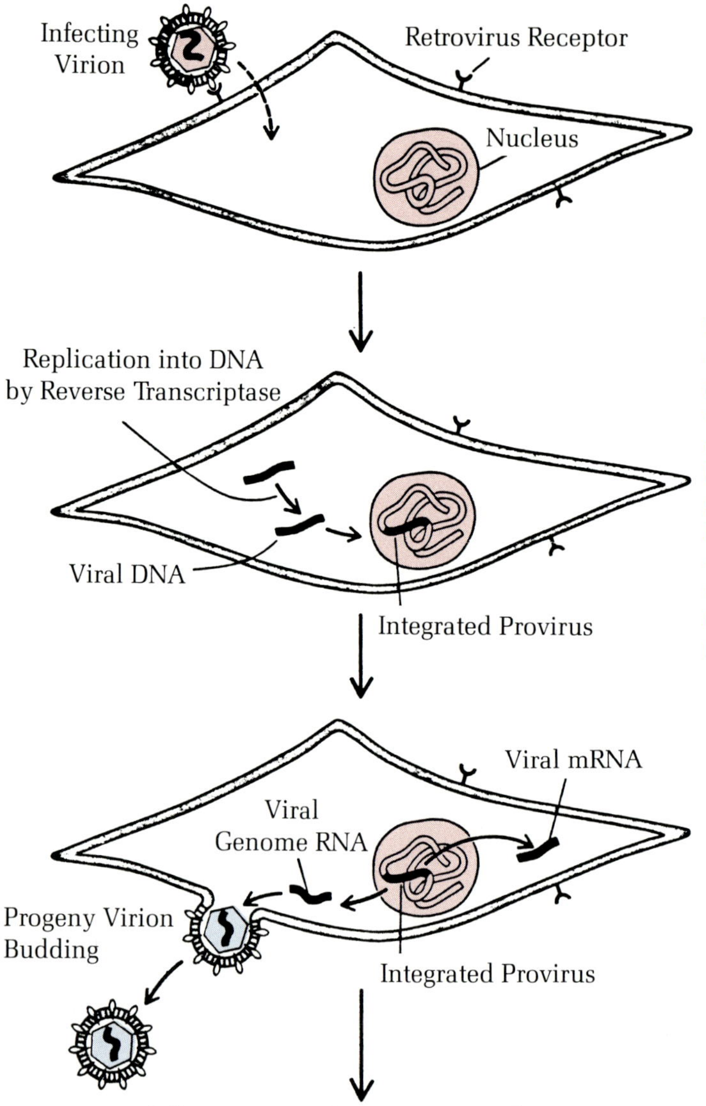

Figure 1. An overview of retrovirus replication. Only one DNA provirus is made per infecting virion; this is replicated by the normal cell division process, following integration into the cell genome. An infected cell is resistant to superinfection by another virus once it starts synthesizing viral glycoprotein, which binds to the receptors and so blocks infection. Usually the cell is not killed.

Acute Transforming Retroviruses. In 1910, long before this virus and its life cycle were understood in detail, Peyton Rous was given a hen with a tumor (Rous, 1911). The hen was derived from a healthy flock of Plymouth Rock chickens from upstate New York. Rous quickly determined that this tumor, termed a sarcoma because it originated in connective tissue, could be transplanted in chickens. Filterable extracts could reproducibly induce an identical tumor near the site of the injection. The Rous sarcoma virus (RSV) induced a reproducible tumor within weeks. And the virus replicated in this tissue, that is, was readily reisolated from the tumor.

A great deal of time passed before assays were developed in cell or tissue culture that permitted the growth and quantitation of RSV. Around 1958 it became clear that RSV infection of chicken cells (connective tissue fibroblasts) in culture dramatically altered the properties of these cells. It produced a transformation in the structure, growth patterns in colonies, and growth abilities in suspension culture of the infected cells (Temin and Rubin, 1958). These changes, collectively called the "transformed phenotype," were clearly a property conferred upon these cells by the virus. Transformed cells produced virus. The next clue (1970) came from experiments, which showed that mutations in the "transforming gene" of RSV could prevent the transformed phenotype, but had no effect upon essential replicative functions (*gag*, *pol*, *env*) of the virus (Martin, 1970). The gene for transformation was not the same as the genes for viral duplication.

By 1976, gene isolation technology had progressed so that the "extra genetic information" in the virus or transforming gene could be isolated and shown to be absent from simple retroviruses composed only of the *gag*-*pol*-*env* genes. Simple retroviruses with only *gag*-*pol*-*env* do not cause sarcomas. RSV was composed of *gag*-*pol*-*env* and a transforming gene (Stehelin et al., 1976). This transforming gene, which came to be called *src* (for sarcomas), was next shown to be closely related to a normal chicken cell gene, based upon DNA nucleotide sequence homology (Stehelin et al., 1976). Somehow, in the chicken, the normal or proto-oncogene was picked up and incorporated into the chromosome of a simple retrovirus (LTR–*gag*-*pol*-*env*–LTR) to produce an LTR–*gag*-*pol*-*env*-*src*–LTR, a rapid-transforming tumor-producing virus. In RSV, the *src* oncogene nucleotide sequence is different but very closely related to its proto-oncogene. The mutations activate the gene and produce a product that causes tumors. We know that *src* is essential for tumor production and transformation because some additional mutations in oncogene *src* eliminate this phenotype. Indeed, a *src* gene clone (in the absence of the viral genes) transforms cells in culture. This result indicates that *src*, not any of the virus components (*gag*, *pol*, *env*), is responsible for the cancer.

This experimental protocol has been repeated some 20–25 times in chickens, turkeys, mice, cats, rats, and monkeys. It yielded retroviruses (some defective for replication because of mutations in *gag*, *pol*, or *env*) with oncogenes that all derive from corresponding proto-oncogenes (Table II). Remarkably, many of the proto-oncogenes fall into families that are related at the genetic or nucleotide sequence levels, with similar proteins and functions. This process will be reviewed in a subsequent section.

Slow Neoplasias of Simple Retroviruses. RSV contains a set of genes (LTR–*gag*–*pol*–*env*–*src*–LTR) that transforms cells in culture and forms tumors within weeks in chickens. *Src* is clearly responsible for these properties. These observations did not explain, however, how simple retroviruses that contained no oncogenes and did not transform cells in culture could produce bursal lymphomas (cancer of the antibody-producing B-cells) with very long latency times (months) in chickens. As more bursal lymphoma tumors were examined and the integration site of the viral DNA copy in the host cell chromosome was examined, a remarkable finding emerged. Most lymphomas have a common integration site, in spite of the fact that integration can take place at random sites. The provirus (integrated viral DNA) was found to be adjacent to a known oncogene termed *myc* (Hayward et al., 1981; Payne et al., 1982).

Thousands, perhaps millions, of retroviruses replicating for a long time in chicken cells had resulted in millions of random integration events. Some of the integration events happened to occur near a proto-oncogene, producing a *pol*–*env*–LTR–

Table II. The Oncogenes

Name	*Origins*	*Tumors*
Protein Kinases		
src	chicken virus	sarcoma
fps/fes	chicken/cat viruses	sarcoma
yes	chicken viruses	sarcoma
ros	chicken viruses	sarcoma
fgr		
abl	mouse/cat viruses	B-cell lymphoma
met	human tumor	
Growth Factors		
sis	wooly monkey/cat viruses	sarcoma (related to PDGF)
Interleukin-III	cloned gene in a retrovirus	myeloid leukemia, from transformed cells
NGF	cloned gene in a retrovirus	(nerve growth factor)
Membrane Receptors (protein kinase activity, tyrosine specific)		
erb B	chicken virus	erythroleukemia and sarcomas (related to the EGF receptor)
fms	cat virus	sarcoma (related to the CSF-1 receptor)
Transmembrane		
neu	human tumor	neuroblastoma (transmembrane protein)
trk	human tumor	tyrosine-protein kinase
ros	tumor	glioblastoma, tyrosine kinase transmembrane
mas	human tumor	transmembrane, rhodopsinlike, β-adrenergic receptorlike
Inner Plasma Membrane (GTPase)		
Ha ras/bas	rat/mouse viruses	sarcoma, erythroleukemia
	human bladder tumor	carcinoma
	mutagenesis-rat	mammary carcinoma
Ki ras	rat virus	sarcoma, erythroleukemia
	human tumor	colon, lung carcinoma
N-ras	human tumor	neuroblastoma, lung carcinoma
Nuclear location		
myc	(a) chicken virus	carcinoma, sarcoma, myelocytoma
(at least	(b) activation by LTR insertion-promotor	Bursal lymphoma-chicken
c-myc, N-myc, L-myc)	(c) activation by chromosome translocation	B-cell lymphoma-mouse, human
myb	chicken virus	myeloblastic leukemia
mos	mouse virus	sarcoma
rel	turkey virus	lymphatic leukemia
ski	chicken virus	sarcoma
fos	mouse virus	sarcoma
p53	virus interactions	bound to SV40 T antigen transforms cells by itself or with ras

myc order of genes in the chicken chromosome. The LTR is a very powerful signal for the high-level production of the adjacent gene *myc m*-RNA and protein. The normal regulation of the *myc* gene is disrupted by the integration event. A B-cell lymphoma gets started in a cell where the *myc* oncogene is activated or overproduces its product, and this is then termed the "promotor insertion" model of leukemogenesis. The leukemogenic cell is selected from among millions of viral DNA integrations by its ability to grow under conditions where normal cells do not replicate. It is remarkable that a totally different mechanism for cancer activation, promotor insertion, uses the same oncogene that was found to be trapped in some retroviruses.

Chromosomal Translocations

Chromosomal abnormalities—such as translocations (fusion of one portion of a chromosome with another chromosome), inversions (a reversal in the order or sequence of genetic information), or deletions (removal of genetic information)—have been detected in malignant cells over a long period of time (Rowley, 1984). The significance of such rearrangements in the genetic information became appreciated, but not necessarily clearer, when it was found that specific leukemias or lymphomas tended to have the same translocation in most or all patients. P. C. Nowell pointed out that almost all patients with chronic myeloid leukemia (CML) carried a translocation involving chromosomes 9 and 22 in the leukemogenic cells (called the Philadelphia chromosome), but not the normal cells (Nowell and Hungerford, 1960). Similarly, Burkett lymphomas commonly contain a translocation of chromosomes 8 and 14; 8–2 and 8–22 translocations have also been observed (Cole et al., 1984; Croce et al., 1984; Hayday et al., 1984).

Molecular clones of DNA were employed to explore the gene organization and composition of DNA at these translocation break points. An oncogene could often be found near the break point, localized to one of the two newly fused chromosomes. In the case of Burkett lymphomas, it was the *myc* oncogene from human chromosome 8. With CML, the *abl* oncogene from chromosome 9 was detected at the break point. In Burkett lymphomas the genetic information donated by the second chromosome involved in the translocation was a portion of the immunoglobulin gene, which was actively being synthesized in these B-cell tumors. The enhancer–promotor signals of the immunoglobulin genes, like those of the retrovirus LTR, were often found at or near the break point and were donated by chromosomes 14, 2, or 22. These enhancer–promoter signals presumably activate high or inappropriate levels of the *myc* proto-oncogene in lymphoblastoid cell lymphomas. The genetic information donated by chromosome 22 to form the Philadelphia chromosome in CML has been termed *bcr*. Its function and genetic composition remain to be determined (Groffen et al., 1984).

Thus, chromosome translocations act like retrovirus "promotor insertions" of an LTR unit. They bring together genetic elements that alter the regulation and activate proto-oncogenes contributing to the formation of cancer. Yet a third mechanism (retrovirus transduction, virus insertion of LTR and translocations of chromosomes) has led experimenters to rediscover the same oncogenes.

Activated Oncogenes from Human Tumors

DNA extracted from cells derived from some human tumors can transform mouse cells in culture if the appropriate genetic information is introduced into these normal cells. If this DNA is extracted from these transformed cells and reintroduced into normal mouse cells, transformed colonies or foci of cells can be detected. The transformed phenotype can be transmitted as a stably inherited trait (Shih et al., 1979). Furthermore, human DNA from normal cells or tissue rarely produces a transformed focus or event. It has been possible to purify and identify the human cell genetic information that can transform mouse cells in culture. This process showed that the *ras* oncogene (first recognized in the Harvey murine sarcoma virus) was responsible for the transformation event (Tabin et al., 1982). Furthermore, the *ras* oncogene differs from its normal human counterpart or proto-oncogene by a single mutation at a specific site in the gene (Blair et al., 1981; Tabin et al., 1982). This mutation event occurred in the normal somatic tissue of the body contributing to the cancer. The mutation is propagated only in the tumor cells. Since 1982, when these experiments were reported, a number of new oncogenes have been detected this way (Table II) (Varmus and Levine, 1983; Vande Woude et al., 1984). Clearly, activation of a proto-oncogene to an oncogene can occur via mutation.

DNA transfection is the fourth distinct experimental approach that has identified members of the same family of oncogenes as the genetic and molecular basis of cancer-causing agents. Normal cellular genes can be activated through increased synthesis or mutation to produce oncogenes, which are involved in cancer pathology.

Functions of Oncogene Products

It is clear that oncogenes can play an important role in the development of a cancer. What protein products do the oncogenes encode, and what are their functions? Some oncogene products, like *v-sis*, are closely related in nucleotide sequence to growth factors like the platelet-derived growth factor (Robbins et al., 1984; Waterfield et al., 1984). This is a protein synthesized in megakaryocytes, stored in platelets, and released to promote cell growth during the healing of a wound. An abnormal growth factor, or the overproduction of a growth factor, could cause a loss of growth regulation. Such growth factors act by binding to specific receptor transmembrane proteins on the cell surface. It may not be surprising then that some oncogenes are closely related to the growth factor receptor proteins (Varmus and Levine, 1983; Vande Woude et al., 1984). Many of these growth factor receptors have a cytoplasmic domain with an associated protein kinase activity. This protein kinase often phosphorylates a target protein specifically at a tyrosine residue. Several other oncogene products are proteins localized in the inner plasma membrane facing the cytoplasm, with tyrosine-specific protein kinase activities (Collett and Erikson, 1978). Some oncogene products localized in the inner plasma membrane hydrolyze GTP to GDP (Shih et al., 1980). In yeast, the *ras* family of proteins is involved in controlling the level of cyclic AMP via adenylcyclase regulation. This in turn, regulates protein kinase levels, metabolic activities, and growth control. In this case, *ras* activity is sensitive to the external nutrient environment and participates in signaling the beginning of a division cycle (Vande Woude et al., 1984). Those protein kinases somehow participate in the transduction of signals into the nucleus of the cell, resulting in altered gene expression and replication of the DNA. Therefore, some oncogene products reside in the cell nucleus, although their functions there are not yet clear (Varmus and Levine, 1983).

In retrospect, it is not surprising that the cancer-causing genetic elements in cells should be related to growth factors, receptors for growth factors, and inner plasma membrane proteins that transduce signals and reflect environmental nutrient status. Proto-oncogenes have evolved to regulate normal cell division processes. Activation of these genes, through increased concentrations or mutation, produces uncontrolled growth.

DNA Tumor Viruses

Retroviruses can cause neoplasia by incorporating an oncogene into their genomes and expressing it, as well as by activating a cellular proto-oncogene. In either case, the cancer-causing gene and product derive from the normal host genome. When the transforming genes of the DNA tumor viruses were first identified, it became clear that no cellular homologs could be detected by either nucleotide sequence comparisons or hybridization tests. Oncogenes from the DNA tumor viruses originate in the evolution of the viruses themselves. Not surprisingly then, many of the viral encoded tumor antigens are essential for viral replication, unlike the *src* gene in RSV. In this case viral gene retention is selected for because it is essential for virus replication, as well as being involved in transformation.

Inoculation of a newborn animal with a DNA tumor virus such as SV40, polyoma, or adenoviruses (Figure 2 shows the structure of an adenovirus) results in infection of cells at the injection site. Within weeks, the infectious virus is lost. Some months later a tumor begins to grow at the site of infection. Unlike the retroviruses, infectious virus is not commonly found in the tumors. However, every tumor cell contains a portion of the viral chromosome integrated into the host cell chromosome (Botchan et al., 1976; Sambrook et al., 1980). In a tumor cell a subset of the viral genes is expressed as m-RNA and proteins. These viral encoded proteins are often recognized as foreign antigens by the immune system of the host, which responds by making antibodies. These antibodies are not sufficient to stop tumor growth, but they react with these proteins. This property makes them useful in identifying the viral proteins synthesized in tumor cells. For this reason, the viral proteins made in tumor cells have come to be called tumor antigens (Tevethia et al., 1980). Viruses with mutations or alterations in their genes for tumor antigens no longer transform cells or produce tumors. This finding indicates that SV40, polyoma, and adenoviruses contain a set of genes responsible for transforming cells in culture and producing tumors in animals (Postel and Levine, 1976; Brockman, 1978; Shenk et al., 1980; Varmus and Levine, 1983; Vande Woude et al., 1984). The names of these genes are reviewed in Table II. In general, one to three viral genes out of the total genome are involved in transforming a cell (Varmus and Levine, 1983).

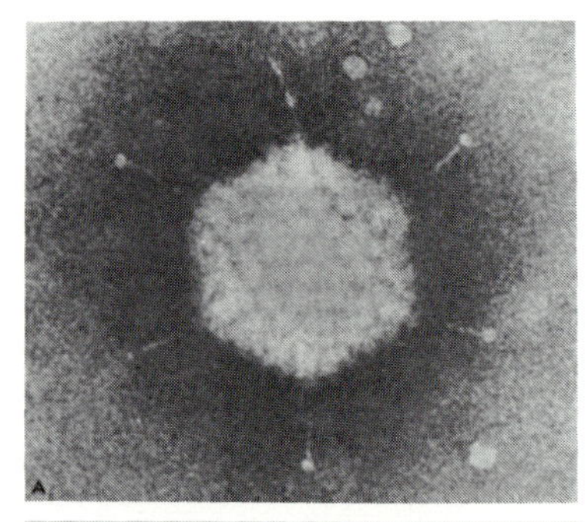

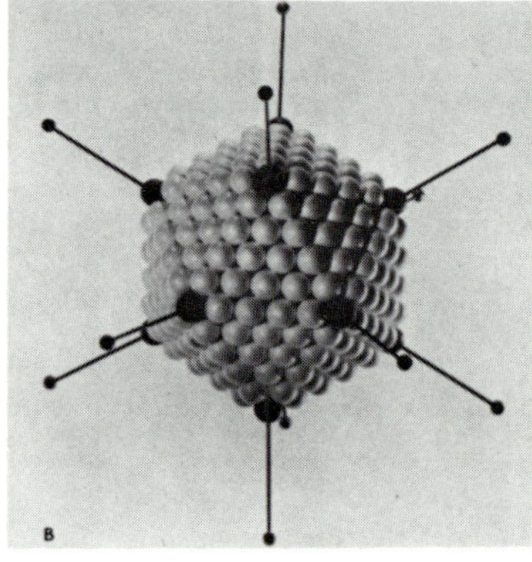

Figure 2. Electron micrograph of human adenovirus type 5 (240,000 ×).

The functions of these viral gene products vary. The adenovirus EIA proteins modulate up or down the levels of transcription of viral and cellular genes (Shenk et al., 1980). The EIB gene products are involved in the transport of messenger RNAs out of the nucleus and into the cytoplasm, where they can be translated into proteins. The SV40 large tumor antigen can stimulate cellular DNA replication in resting cells and promote cell division (Postel and Levine, 1976). It is localized both in the cell nucleus and at the outer plasma membrane (Tevethia et al., 1980). The Epstein–Barr virus LMP antigen, which transforms cells in culture, is localized in the membrane of the cell. EBNA-II protein has a nuclear location. The E5 and E6 proteins of papilloma viruses implicated in cellular transformation and some human cancers have been localized in the plasma membrane, nuclear membrane, and nuclear fractions (Varmus and Levine, 1983).

These viral transforming genes might stimulate the production of cellular proto-oncogenes or modify the proto-oncogene products in the cell, making them oncogenes. The large T antigen of SV40 is found in an oligomic protein complex with a cellular protein p53, which is known to have oncogene activity (Lane and Crawford, 1979; Linzer and Levine, 1979). Similarly, the polyoma middle T antigen is complexed with normal cellular *src* at the inner plasma membrane in a transformed cell (Courtneidge and Smith, 1983). These are two examples of how viral oncogenes act by modifying cellular proto-oncogene products. For example, when the SV40 large T antigen complexes with the cellular p53 protein, it greatly increases the stability of the protein. The half-life of monomeric p53 is about 20 minutes, but T–p53 complexes have half-lives longer than 40 hours (Oren et al., 1981). This stability results in greatly increased levels of p53 in an SV40-transformed cell. Similarly, the normal *c-src* kinase activity is enhanced by the binding of polyoma middle T antigen, and its specificity might be altered as well. Some viral oncogenes derived from the DNA viruses act to alter cellular functions that have been identified as proto-oncogenes. This fact further strengthens the theory that there is a unity of mechanisms involved in cancer causation. Again, the study of the molecular mechanisms of cancer leads to the same family of oncogenes.

Infections of humans with hepatitis B virus can result in long-term chronic virus production and liver damage. These patients are at significantly higher risk to develop liver carcinomas. In this case no viral oncogene has been isolated, and the mechanism of tumor formation remains unknown. The papilloma viruses, closely associated with cervical carcinomas in humans, encode their own oncogene products. These have been identified in tumors and transformed cells (Table I).

Finally, some of the pox viruses (related to the smallpox virus) can induce benign tumors in animals. These viruses seem to contain a gene that encodes the information for a growth factor related to the host cell's epidermal growth factor. The viral growth factor binds to the EGF receptor and apparently triggers it to signal for cell division to begin. In this case a viral oncogene acts upon a cellular target that is itself a cellular proto-oncogene, starting a series of events leading to uncontrolled growth. Yet another experimental approach and observation leads us back to a common theme: the oncogenes.

Conclusions

The study of viruses that are associated with neoplasia has provided fundamental insights and experimental approaches in understanding this disease. Almost all of the DNA virus group (List I) have representatives that can transform cells in culture or produce tumors in animals under the appropriate circumstances. They contain a subset of genes that can alter the growth properties of the host cell. These genes encode proteins, termed tumor antigens, localized in the nucleus or at the plasma membrane. Such gene products function to alter cellular gene expression and promote cell division. In some cases, these viral proteins can be shown to interact with and increase the levels of normal cellular proteins, which themselves are proto-oncogene products (Table I). Of all the groups of RNA viruses (List I), only the retroviruses can cause neoplasia. They do so by either incorporating and altering a cellular gene or by integration of a transcriptionally active element next to a proto-oncogene. Some 20–25 oncogenes appear to be involved in cancers, whether produced by viruses or induced by mutations. The normal function of these proto-oncogene products is to regulate cell growth. The proto-oncogenes encode the genetic information for polypeptides that are growth factors, cellular receptors of growth factors, inner plasma membrane protein kinases that transmit signals within a cell, and nuclear proteins involved in gene regulation and/or the replication of the genome. Activation of these genes, either by altering the levels of normal products or by mutations in the structural genes, changes the qualitative nature of these proteins and produces a product that contributes to causing a cancer. Cancer cell functions are qualitatively and/or quantitatively different from those of normal cells. This is an optimistic note for looking ahead or we can exploit these differences, we may be able to stop this uncontrolled growth.

References

Blair, D. G.; Oskarsson, M.; Wood, T. G.; McClements, W. L.; Fischinger, P. J.; Vande Woude, G. F. *Science (Washington, DC)* **1981,** *212,* 941–943.

Botchan, M.; Topp, W.; Sambrook, J. *Cell* **1976,** 9, 269–287.

Brockman, W. W. *J. Virol.* **1978,** *25(3),* 860–870.

Cancer Cells; Vande Woude, G. F.; Levine, A. J.; Topp, W. C.; Watson, J. D., Eds.; Cold Spring Harbor Laboratory: Cold Spring Harbor, 1984; Vol. 2.

Cole, M. D.; Piccoli, S. P.; Keath, E. J.; Caimi, P.; Kelekar, A. *Cancer Cells* **1984,** *2,* 227–233.

Collett, M. S.; Erikson, R. L. *Proc. Natl. Acad. Sci. USA* **1978,** *75(4),* 2021–2024.

Courtneidge, S. A.; Smith, A. E. *Nature (London)* **1983,** *303,* 435.

Croce, C. M.; Erikson, J.; Nishikura, K.; ar-Rushdi, A.; Giallongo, A.; Rovera, G.; Finan, J.; Nowell, P. C. *Cancer Cells* **1984,** *2,* 235–242.

Microbiology; Davis, B. D.; Dulbecco, R.; Eisen, H. N.; Ginsberg, H. S. Eds.; Third Edition, Harper & Row: New York, 1980.

Groffen, J.; Heisterkamp, N.; Stephenson, J. R.; Grosveld, G.; de Klein, A. *Cancer Cells* **1984,** *2,* 261–272.

Hayday, A. D.; Saito, H.; Wood, C.; Gillies, S. D. Tonegawa, S.; Wiman, K.; Hayward, W. S. *Cancer Cells* **1984,** *2,* 243–252.

Hayward, W. S.; Neel, B. G.; Astrin, S. M. *Nature (London)* **1981,** *290(5806),* 475–480.

Lane, D. P.; Crawford, L. V. *Nature (London)* **1979,** *278(5701),* 261–263.

Linzer, D. I. H.; Levine, A. J. *Cell* **1979,** *17,* 337–346.
Martin, G. S. *Nature (London)* **1970,** *227,* 1021–1023.
Nowell, P. C.; Hungerford, D. A. *Science (Washington, DC)* **1960,** *132,* 1497–1500.
Oren, M.; Maltzman, W.; Levine, A. J. *Mol. Cell Biol.* **1981,** *1,* 101–110.
Payne, G. S.; Bishop, J. M.; Varmus, H. E. *Nature (London)* **1982,** *295(5846),* 209–217.
Postel, E. H.; Levine, A. J. *Virology* **1976,** *73,* 206–215.
Readings in Tumor Virology; Varmus, H.; Levine, A. J., Eds.; Cold Spring Harbor Laboratory: Cold Spring Harbor, 1983.
Robbins, K. C.; Antoniades, H. N.; Devare, S. G.; Hunkapiller, M. W.; Aaronson, S. A. *Cancer Cells* **1984,** *1,* 35–42.
Rous, P. *J. Am. Med. Assoc.* **1911,** *56,* 198.
Rowley, J. D. *Cancer Cells* **1984,** *2,* 221–226.
Sambrook, J.; Greene, R.; Stringer, J.; Mitchison, T.; Hu, S.-L.; Botchan, M. *Cold Spring Harbor Symp. Quant. Biol.* **1980,** *XLIV,* 569–584.
Shenk, T.; Jones, N.; Colby, W.; Fowlkes, D. *Cold Spring Harbor Symp. Quant. Biol.* **1980,** *XLIV,* 367–375.
Shih, C.; Shilo, B.-Z.; Goldfarb, M. P.; Dannenberg, A.; Weinberg, R. A. *Proc. Natl. Acad. Sci. USA* **1979,** *76(11),* 5714–4718.
Shih, T. Y.; Papageorge, A. G.; Stokes, P. E.; Weeks, M. O.; Scolnick, E. M. *Nature (London)* **1980,** *287,* 686–691.
Shope, R. E. *J. Exp. Med.* **1933,** *58,* 607–610.
Stehelin, D.; Varmus, H. E.; Bishop, J. M.; Vogt, P. K. *Nature (London)* **1976,** *260,* 170–173.
Tabin, C. J.; Bradley, S. M.; Bargman, C. I.; Weinberg, R. A.; Papageorge, A. G.; Scolnick, E. M.; Dhar, R.; Lowy, D. R.; Chang, E. H. *Nature (London)* **1982, 30,** 143–149.
Temin, H. M.; Rubin, H. *Virology* **1958,** *6,* 669–688.
Tevethia, S. S.; Greenfield, R. S.; Flyer, D. C.; Tevethia, M. J. *Cold Spring Harbor Symp. Quant. Biol.* **1980,** *XLIV,* 235–242.
Waterfield, M. D.; Scrace, G. T.; Whittle, N.; Stockwell, P. A.; Stroobant, P.; Johnsson, A.; Wasteson, A.; Westermark, B.; Heldin, C.-H.; Huang, J. S.; Deuel, T. F. *Cancer Cells* **1984,** *1,* 25–33.

CHAPTER 5 Genetic Basis of Cancer and Its Suppression

Ruth Sager

The human organism is a tightly controlled system with built-in checks and balances to maintain the body's status quo. The ancient Chinese viewed all nature, including human beings, as a harmonious entity kept in balance by the opposite but interacting forces of Yang and Yin. In cancer this harmonious balance is disrupted at the level of the cell, the tissue, and the organism. How does this disruption come about? How can we restore the natural balance and use it for cancer therapy and prevention?

Genetic Basis of Cancer

Cancer begins as a genetic defect. Later it takes many forms, depending on which tissue of the body has been affected. For example, lung cancer starts with genetic changes in a single cell in the lung. Similarly, cancer of other organs, such as the breast, colon, ovary, or prostate, begins with a single cell in those organs. Leukemia starts with genetic changes in the bone marrow, where white blood cells are

1420–4/88/0075$06.50/0

produced. However the disease develops in its later stages, it begins as gene mutations and chromosomal rearrangements (Rowley and Ultmann, 1983).

Some of these genetic changes, inherited from generation to generation through the egg or sperm, are called *germinal*. However, the diseased condition is not inherited; individuals who develop cancer later in life are initially healthy. What is inherited is an increased chance or probability that cancer will occur.

We call this condition an inherited predisposition, and it is often associated with a particular kind of cancer. For example, in the inherited disease retinoblastoma, the probability is high that anyone born with a mutated or deleted retinoblastoma gene (called *Rb*) will have tumors in the retina of the eye as an infant (Knudson, 1985). Fortunately, many of these tumors are operable and lives can be saved by prompt surgery.

In retinoblastoma, as in all other examples of genetic predisposition to cancer that have been studied, the inherited defective gene does not act alone. For tumors to arise and grow, other genetic changes (not transmitted germinally to offspring) must also occur. These other changes are *somatic* (in contrast to germinal). All cells of the body are somatic except the germ cells, that is, eggs and sperm. Somatic mutations occur during development or in adults, and their frequency can be greatly increased by environmental factors such as radiation and chemical carcinogens. Thus, individuals may or may not carry genes that confer an inherited predisposition. In either case, the frequency of somatic mutations governs the origin and progression of the disease.

How do we detect inherited predisposition to cancer? The classical method is pedigree (family tree) analysis. Family trees have been studied by observant physicians and others since prehistoric times.

An example of a retinoblastoma pedigree is shown in Figure 1 (Sparkes et al., 1983). The inheritance of retinoblastoma is depicted, together with a closely linked gene, esterase D, for which two alleles (here labeled 1 and 2) are known. The black symbols indicate persons with retinoblastoma. The disease was transmitted through four generations, always linked with esterase B allele 2. There is no evidence of sex preference. Thus the pattern of inheritance is that of an autosomal dominant gene. Other studies have located the *Rb* gene on the long arm of chromosome 13, that is, chromosome 13q.

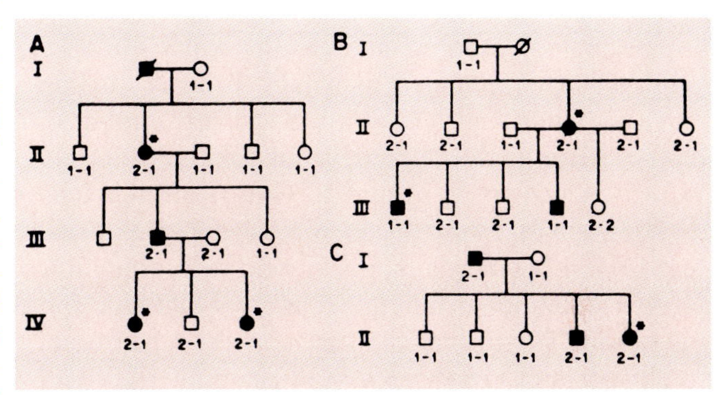

Figure 1. Retinoblastoma family pedigree. The darkened symbols indicate persons with retinoblastoma; an asterisk (*) by the symbol indicates that the tumor is unilateral. The tumor developed early in the affected persons' childhood and had a typical presentation for retinoblastoma. The numbers under the individual symbols indicate the esterase D genotype. (Reproduced with permission from Sparkes et al. Copyright 1983 American Association for the Advancement of Science.)

Analysis of retinoblastoma has revealed the importance of gene loss in cancer. Pedigree analyses show a consistent correlation between loss or mutation of the *Rb* gene and expression of the disease. However, in the inherited form of the disease, only one copy of the gene is missing (or inactive) from all the normal cells in individuals who are carriers and transmit the disease to progeny. They are heterozygous for the *Rb* gene. When tumors from these same individuals are examined, BOTH copies are found to be absent (or inactive) in the tumor cells. Thus, the tumor cells are homozygous deficient. This result shows that one copy of the normal *Rb* gene is sufficient to protect the individual, maintaining normality and health. Tumor formation involves loss of both copies of the gene and therefore of the protein encoded by the gene. The genetics of *Rb* is discussed in two recent reviews (Knudson, 1985; Murphree and Benedict, 1984). The role of genes like *Rb* in tumor suppression will be examined later, after a look at how the loss of the *Rb* genes is detected experimentally.

The kinds of chromosome changes that occur in somatic cells, leading to loss of a gene or a larger region of a chromosome, are shown in Figure 2 (Murphree and Benedict, 1984). One entire chromosome 13, carrying the normal *Rb*$^+$ gene, may be lost by nondisjunction. This loss leaves the progenitor cell that gives rise to the tumor permanently missing that chromosome. In nondisjunction at mitosis, the two copies of a newly replicated chromosome are not separated at metaphase, and both copies go to one daughter cell. The homologous chromosome separates normally. In consequence, one daughter cell gets three copies and the other gets only one. Sometimes the single chromosome gets reduplicated at a later mitosis, so that each daughter cell recovers the correct number of copies. Each is now homozygous for that chromosome, as shown in Figure 2. Other chromosome aberrations that may produce a homozygous product from a heterozygote are mitotic recombination, a new deletion of the *Rb* gene, inactivation, and mutation. These changes can convert a germinal predisposition to cancer into the somatic reality of tumor formation. However, most of these changes cannot be detected easily in the microscope. To follow the changes that occur, it is necessary to use molecular techniques that examine DNA molecules rather than entire chromosomes.

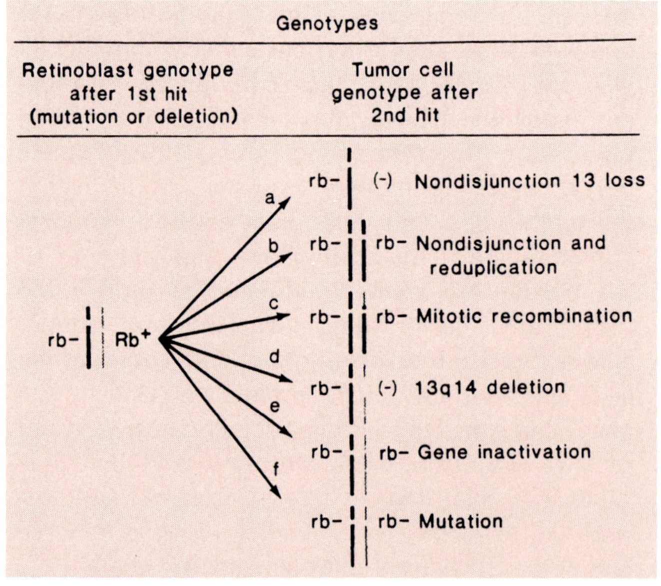

Figure 2. Several mechanisms that would produce hemizygosity or homozygosity at the Rb locus are depicted. The "first hit", which could have occurred previously at the germinal level or independently at the somatic level within the retinoblast, is shown at the left. It could include a point or frameshift mutation that could not be detected microscopically, or a deletion that could be detected microscopically resulting in the inactivation of one Rb allele (rb— or —). A loss of the normal Rb allele might then occur as the second event leading to hemizygosity at the Rb locus (a) in which only the inactivated or deleted allele remained. A nondisjunctional loss as the second event could be followed by reduplication of the remaining rb— allele (b), producing homozygosity at the Rb locus. A mitotic recombination that includes the rb— allele in the recombination (c) is also a possibility, as is a microscopic deletion or a submicroscopic deletion (d). Inactivation of the remaining Rb+ allele can also be a rare occurrence resulting from the translocation of 13q14 onto an inactive X chromosome (e). In such a case, the structural Rb+ allele would still be present but would not be functional. Finally, the second event might be a point or frameshift mutation of the remaining Rb allele (f). (Reproduced with permission from Murphree and Benedict. Copyright 1984 American Association for the Advancement of Science.)

A very powerful new method is based on the recent discovery that we all carry simple mutations in our DNA that cause no ill effects: so-called neutral mutations. They are often single-base changes: a T (thymine) for a G (guanine) or A (adenine), for example. This substitution is recognized by a class of restriction enzymes that are each able to bind to a particular series of bases in DNA and, after binding, to cleave the double-stranded DNA at that location. For example, the restriction enzyme called EcoRI recognizes the sequence GAATTC and will cut at that site, as shown in Figure 3, but will not cut if a single base has been changed. Any site in DNA at which a base change has occurred in some individuals but not others is called a DNA polymorphism.

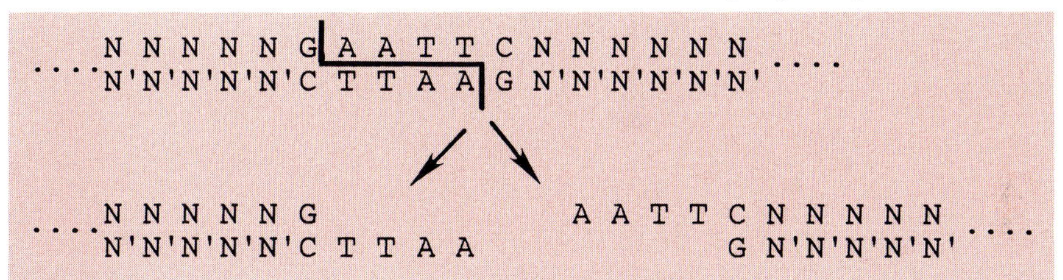

Figure 3. The restriction enzyme EcoRI recognizes the restriction site $\frac{\text{GAATTC}}{\text{CTTAAG}}$ in any DNA sequence. Here we show N and N′ as complementary bases of a long DNA molecule in which the EcoRI site is embedded. The enzyme makes a staggered cut in the DNA, producing two DNA molecules, each with a characteristic staggered end that can be used in genetic engineering.

Since we each have thousands of inherited polymorphisms, it is evident that the chromosome pairs that we inherited are not identical (Botstein et al., 1980). Sometimes different nucleotides are present at a particular location on one chromosome and its homolog. Detection of such polymorphisms is a fool-proof way to distinguish which chromosome came from which parent, provided a DNA sample from each parent is available. Because the genetic constitution of all of one's somatic cells is identical (except for rare mutations), any cells, such as skin or white blood cells, can be used to provide a DNA sample.

Restriction fragment length polymorphisms (RFLPs) are being used to detect inherited predispositions to cancer and somatic mutation by RFLP-pedigree analysis, as shown in Figure 4 (Cavenee et al., 1985). Here we compare DNA from normal cells of a patient with DNA from the tumor of the same patient. We look for a change in restriction fragment pattern to show that a somatic change has occurred.

In this simple pedigree, we examine two generations: the parents and one male offspring. Both the father and son had retinoblastoma. Four restriction site polymorphisms (HU10, 9D11, 1E8, and HUB8)

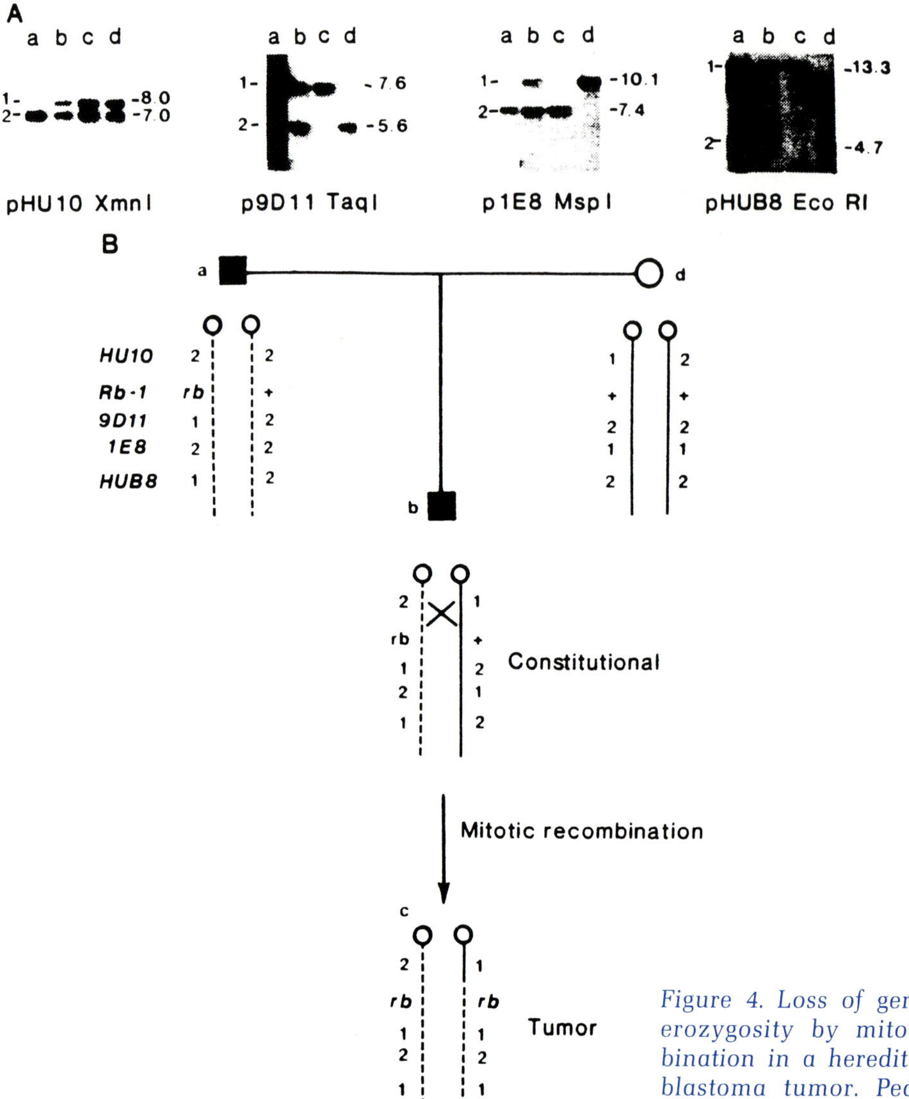

Figure 4. Loss of germline heterozygosity by mitotic recombination in a hereditary retinoblastoma tumor. Pedigree and inferred chromosomal haplotypes. Filled symbols represent individuals with retinoblastoma. The × indicates the probable point of mitotic crossover. (Reproduced with permission from Cavenee et al. Copyright 1985 American Association for the Advancement of Science.)

that had previously been mapped in relation to *Rb*-1 were used as shown in Figure 4B. The father, a, was heterozygous for the *Rb* gene (+ means normal; *rb* means mutant), but his tumor cells were not examined to find the somatic mutation that resulted in formation of his retinal tumors. Of course, his somatic mutation was not inherited; the son received only germinal chromosomes. Numbers 1 and 2 refer to different polymorphic alleles recognized by the restriction enzymes, whose DNA cleavage effects are shown in Figure 4A. We see the cleavage patterns for each enzyme (XmnI, TaqI,

MspI, and EcoRI). DNAs, in separate lanes, are derived from the two parents, a and d; from unaffected cells of the son, b; and from his tumor, c. By comparing the patterns, we can infer that the somatic change that led from the heterozygous condition, *rb*/+, to the tumor condition, *rb*/*rb*, was probably a mitotic recombination, as shown in the figure. (In Figure 1, *rb*/+ is referred to as *Rb*+/*rb*− and *rb*/*rb* as *rb*−/*rb*−.)

How do we know where to look in the genome for DNA changes? Which region of which chromosome shall we examine? Previous microscopic studies of chromosomes associated with retinoblastoma had revealed suspicious deletions in the long arm, q, of chromosome 13, that is, in 13q. Other linked genes such as esterase D (Figure 1) and RFLPs known to be located on 13q were then used to confirm the occurrence of chromosome changes in the region of the *Rb* gene, in cells derived from retinal tumors.

When the approximate location of a gene associated with the particular cancer is not known, RFLP analysis is a trial and error procedure. In malignant melanoma, for example, RFLP analysis has identified deletions on almost every chromosome. We do not yet know which of these aberrations are carcinogenic. On the other hand, similar studies of breast cancer have already identified several sites on different chromosomes that show some correlation with the disease. Pedigree studies of DNA will be necessary to determine which changes are germinal and which are somatic. This potentially powerful method is still in an early stage of application.

Cancer as a Genetic Disease

The idea that cancer is caused by chromosomal changes was first proposed in a classical book by Boveri. Entitled *The Origin of Malignant Tumors*, it was published in German in 1914 and in English translation in 1929 (Boveri, 1929). However, the hypothesis was difficult to establish with methods available in the early 1920s. This view has gained wide acceptance since the late 1960s, when new methods were developed for chromosome and DNA analysis. It is worthwhile to review very briefly the chief lines of evidence that cancer does originate as a disease of the genome.

• Solid tumors and leukemias are clonal in origin (Nowell, 1976), as shown diagrammatically in Figure 5. This means that cancers arise from single malignant cells that go on to multiply and produce clones. Cancerous events are transmitted from a cell to its many progeny during somatic growth and development. Clonality has been demonstrated principally by examining specific chromosomal changes, as well as mutations that can be used to distinguish the clone from adjacent cells of a different origin.

One implication of clonality is that the incidence of cancer at the cellular level is exceedingly rare. We have about 10^{14} cells in our body, of which about 1% could give rise to cancer. (Fully differentiated cells—such as bone, muscle, nerves, and red blood cells—are incapable of further cell division and, therefore, never form tumors.) Nearly all of us have some form of cancer at some time in our lives. The probability per individual is near 1, but the probability per cell is only about 1 in 10^{12}.

In fact, evidence suggests that we each have many microscopic premalignant lesions that never progress to a detectable stage (Cairns, 1978). This means that the initial incidence may be much higher than 1 in 10^{12}. If so, there must be other mechanisms that block tumor growth. We already know several examples of genes that encode growth inhibitors. The suppression of tumor growth will be discussed in the last section of this chapter.

• The fact that X-ray exposure increases the frequency of cancer was one of the first lines of evidence for the role of external agents in carcinogenesis. More recently ultraviolet radiation and a whole host of chemicals have also been implicated as carcinogens. Many lines of investigation have shown that all these agents act by inducing permanent changes in DNA.

• Cytogenetic analysis has shown direct association of particular kinds of cancer with specific

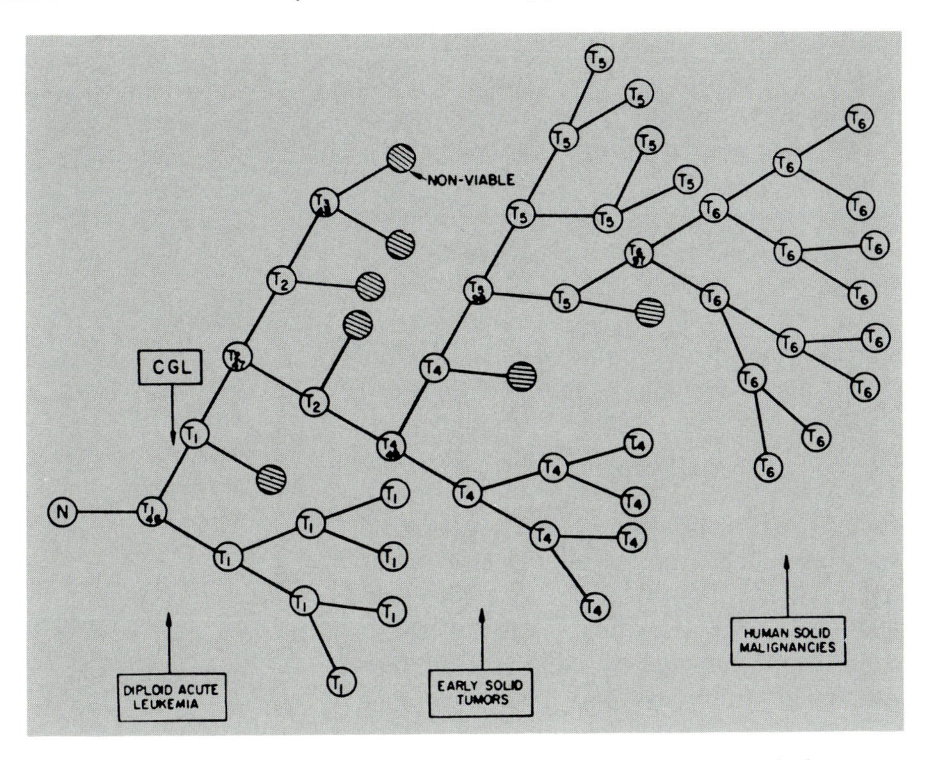

Figure 5. Model of clonal evolution in neoplasia. Carcinogen-induced change in progenitor normal cell (N) produces a diploid tumor cell (T_1, 46 chromosomes), with growth advantage permitting clonal expansion to begin. Genetic instability of T_1 cells leads to production of variants (illustrated by changes in chromosome number, T_2–T_6). Most variants die, due to metabolic or immunologic disadvantage (hatched circles). Occasionally one has an additional selective advantage (for example, T_2, 47 chromosomes), and its progeny become the predominant subpopulation until an even more favorable variant appears (for example, T_4). The stepwise sequence in each tumor differs, being partially determined by environmental pressures on selection. This variation results in a different aneuploid karyotype predominating in each fully developed malignancy (T_6). Earlier subpopulations (for example, T_1, T_4, T_5) may persist sufficiently to contribute to heterogeneity within the advanced tumor. Biological characteristics of tumor progression (for example, morphological and metabolic loss of differentiation, invasion and metastasis, resistance to therapy) parallel the stages of genetic evolution. Human tumors with minimal chromosome change (diploid acute leukemia, chronic granulocytic leukemia) are considered to be early in clonal evolution. Human solid cancers, typically highly aneuploid, are viewed as late in the early developmental process. (Reproduced with permission from Nowell. Copyright 1976 American Association for the Advancement of Science.)

chromosomal abnormalities. Many examples are given in Rowley and Ultmann, 1983. Retinoblastoma and associated changes in chromosome 13 were discussed already. The somatic translocation t(9;22) has been found in bone marrow precursor cells in a very high percentage of patients with chronic myelocytic leukemia (CML), and t(8;14) has been clearly associated with Burkett's lymphoma. The nonmalignant cells of these patients have the normal chromosome complement. By now, at least 100 examples of cytogenetic correlations with specific forms of cancer have been described.

In summary of what has been presented so far, cancer is a genetic disease with both germinal and somatic aspects. We may inherit a predisposition to cancer, but the development of the disease depends on somatic mutations and rearrangements in DNA that occur later during development or in adults. Thus, genetic predisposition increases the likelihood of cancer, but the extent of the increased probability depends on the particular genes involved.

Acceptance of the genetic basis of cancer gained increasing momentum in the late 1970s and early 1980s. The past few years have seen an overwhelming amount of evidence based on the identification of specific genes with determining effects on tumor formation. These genes can be divided into three classes: (1) genes that affect DNA synthesis and repair, in which mutations lead to production of defective DNA; (2) genes whose mutations promote tumor formation; and (3) genes that suppress or inhibit tumor formation.

Genes That Affect DNA Synthesis and Repair

A series of inherited diseases with complex symptoms, including the production of cells with chromosome breaks and rearrangement, as well as a high probability of cancer, have been known for many years (German, 1983). They include xeroderma pigmentosum, in which patients are very susceptible to skin cancer induced by exposure to sunlight; Bloom's syndrome, in which individuals are stunted in growth and very susceptible to infectious diseases as well as to cancer; and ataxia telangiectasia, Fanconi's anemia, and Werner's syndrome, all of which involve chromosomal instability. In xeroderma, the gene mutation results in defective DNA repair of lesions induced by the UV component of sunlight. The enzymatic basis is not understood in the other diseases, except for Bloom's syndrome. There it has recently been shown that the defect is in DNA ligase, a critical enzyme required for normal DNA synthesis and repair (Willis and Lindahl, 1987). Historically, these diseases have played very important roles as they focussed attention on DNA aberrations associated with cancer, both because they are inherited diseases and because the inherited mutations lead directly to the DNA breaks and rearrangements seen in the tumors that develop.

Genes That Promote Tumor Formation

The discovery of a set of genes that promote cancer (oncogenes) and of genes that inhibit cancer (tumor-suppressor genes) has provided the most insightful information yet available on the origin and progression of cancer.

Mutations that alter structure or regulation of specific genes might be tumor-inducing. This is an old idea, but the identification and cloning of genes with this property has just occurred within the past few years. Two kinds of experiments led to this success (Varmus, 1984). One was the analysis of viral nucleic acid sequences that induce transforming effects when introduced into certain cell lines; the other was the introduction of DNA from tumor cells into non-tumor-forming cells. Rare cells that appeared aberrant could be recovered and grown in culture to form large populations of cells. Then the specific genes with tumor-inducing effects could be cloned out. Both of these experimental approaches have been highly productive. The tumor-inducing genes identified in these experiments, called transforming genes or oncogenes, are discussed in Chapter 4.

Oncogenes have been very important in demonstrating the effects of individual gene mutations and altered gene expression in tumorigenesis. Although their detailed molecular activity remains unclear, oncogenes have provided us with a set of previously unknown proteins that play a central role in the control of growth and differentiation. To discover new genes of this type, scientists are studying newly isolated viruses, transferring DNA from tumor cells into normal cells, and selecting for the rare aberrant or tumor-forming cell that has incorporated transforming DNA. The DNA transfer methods used in the quest for new oncogenes are also the methods of choice in the search for genes that suppress tumor formation.

Genes That Suppress Tumor Formation

The existence of tumor-suppressor genes has been shown most clearly with two kinds of evidence. First, in certain hereditary cancers both copies of a single locus have to be lost or inactivated for tumors to arise. Thus, by inference, this (normal) region encodes a suppressor of tumor formation (Knudson, 1985; Sager, 1986).

The example of retinoblastoma will help to clarify this inference. In the hereditary form of the disease, one copy of the *rb* gene is absent or inactive, and the homologous gene on the other chromosome 13 is presumed to be normal. Thus, the patient is heterozygous for the *Rb* gene in all of the somatic cells of the body. In the tumor, however, some changes have arisen somatically. The tumor cells differ from the normal cells of the patient, as shown in Figure 4, by the loss of a region of the chromosome around the *Rb* gene. The polymorphic marker genes have gone from heterozygous to homozygous. This change is associated with loss of the normal gene, as shown in the RFLP-pedigree analysis.

Thus, in order for the tumor to arise, both copies of the normal *Rb* gene have to be inactive. This loss leads to the inference that the normal gene encodes a gene product or protein that inhibits tumor formation. To establish this inference as fact, it will be necessary to transfer the *Rb* gene into a retinoblastoma tumor cell in order to directly demonstrate its tumor-suppressing effect. Several laboratories are working on this project. The *Rb* gene is not unique as a suppressor of tumor formation. A number of genes that must be deleted for specific tumors to arise have now been identified, and their number is growing.

One example is Wilms' tumor (Rowley and Ultmann, 1983, Chapter 6). The gene involved has been located on the short arm of chromosome 11. When the disease is inherited, the somatic cells are heterozygous for a deletion in this chromosomal region; in the tumor cells themselves, the deletion has become homozygous, occurring on both copies of chromosome 11p (Housman et al., 1986). RFLP-pedigree analysis has also shown that a complex of developmental anomalies, called the Beckwith–Wiedemann syndrome, are also associated with homozygosity in 11p (Koufos et al., 1985).

A deletion in chromosome 3p has recently been associated with small-cell lung cancer (Brauch et al., 1987), and a deletion in chromosome 5q with a hereditary form of colon cancer (Bodmer et al., 1987; Solomon et al., 1987). Deletions in chromosome 22 have been associated with certain tumors of neural origin, meningiomas, and acoustic neuromas, as well as malignancies developing in neurofibromatosis (the Elephant Man's disease) (Seizinger et al., 1986), as suggested earlier by Knudson (1985).

Tumor-suppressor genes have also been inferred from cell fusion experiments. The basic procedure is to fuse pairs of cells, one normal and the other tumorigenic or tumor-derived (N × Tu fusions), as shown in Figure 6. Hybrid cells are recovered by selection, using genetic markers present in the parental cells. After the hybrids are grown up to several million cells, they are reinjected into animals to test for tumor formation.

The remarkable and unexpected result of this experiment is that the hybrids are usually NOT tumorigenic. In other words, some gene product(s) from the normal cells, expressed in the hybrid cells, is (are) strong enough to inhibit the tumor-forming ability of the tumor parent.

The N × Tu fusion experiment and subsequent analysis of hybrid clones has been carried out by many investigators using different combinations of human and rodent N and Tu cells (Sager, 1986). In general, tumor suppression is clearly evident in hybrid clones. One complication in this kind of experiment is that some chromosomes are lost from the hybrid cells, often at random. Loss of the chromosome carrying the suppressor gene will, of course, lead to loss of suppression—that is, to the restoration of tumor-forming ability. If this occurs, the suppressed cells will give rise to tumorigenic daughter cells.

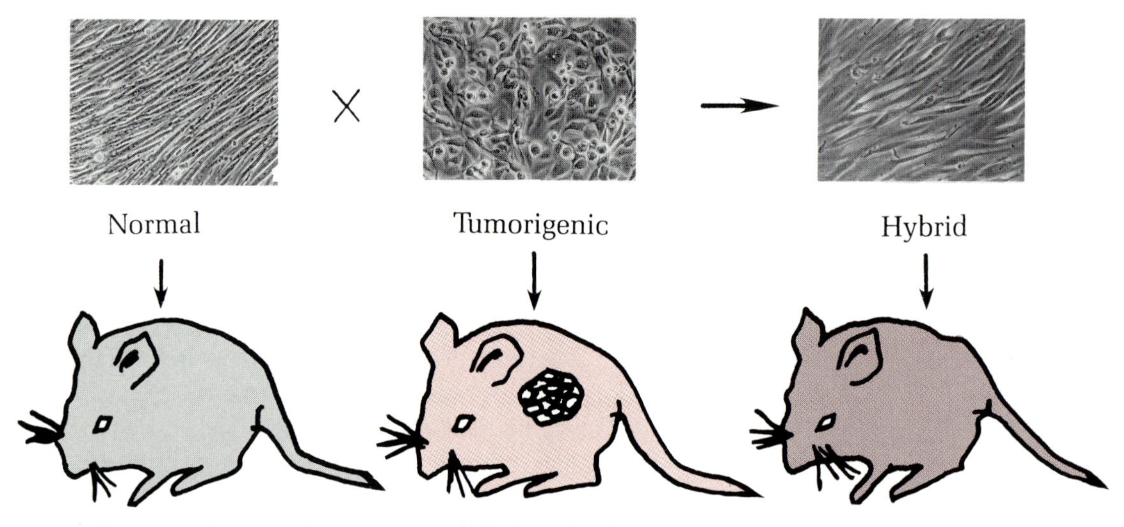

Figure 6. Suppression of tumor formation in cell hybrids of normal x tumor cells. Here, normal cells were fused with tumorigenic cells to form hybrids that resemble the normal parent. When these cells are grown to millions of cells and injected into nude mice, they do not form tumors.

This result can be very useful for further analysis. The resulting tumor cells can be analyzed to determine which chromosome is lost in tumorigenic hybrids but retained in nontumorigenic hybrids. By this means, a few chromosomes have been identified as potential carriers of suppressor genes: chromosome 4 in the mouse, chromosome 15 in the Syrian hamster, chromosome 3 in the Chinese hamster, and chromosome 11 in humans.

Chromosome 11 has also been implicated by a different method as carrying a suppressor of human Wilms' tumor disease. A human chromosome containing the 11p region was introduced into Wilms' tumor cells by means of microcell transfer (Weissman et al., 1987). In this technique mitotic cells are treated with the drug cytochalasin B, which induces formation of micronuclei containing one or a few chromosomes. Those cells that received the Wilms' tumor region on 11p were selected by a linked genetic marker and grown for tumor testing. The Wilms' tumor cells that previously made tumors in nude mice were unable to produce tumors when they contained the transferred chromosome with its normal Wilms' tumor gene. This inactivity demonstrated the tumor-suppressing effect of the chromosome.

Thus both lines of evidence, chromosomal localization of genes whose loss predisposes individuals to cancer and tumor suppression in cell hybrids from N × Tu fusions, provide means to identify tumor-suppressor genes. Within the next few years many of these genes will probably be cloned and characterized. It will then be possible to investigate their interaction with the oncogenes that promote tumor formation. At present we can make some reasonable inferences about how tumor-suppressor genes might act.

Possible Functions of Tumor-Suppressor Genes

Direct Reversal of Oncogene Action. Some oncogenes are mutated forms of normal proto-oncogenes that function in normal growth control. Therefore, one possible way to reverse oncogene action would be to introduce more copies of the normal gene into cells.

Genes That Encode Growth Inhibitors. The loss of growth regulation is a central feature of tumorigenesis. Normal cells are tightly controlled as to the conditions under which cell division is permitted. In wound healing, for example, cells grow rapidly to repair the wound. However, they also stop growing in a precise way, so that the wound does not produce a tumor. Most of the oncogenes, genes that promote tumor formation (discussed in Chapter 4), act by inducing cell proliferation at the wrong time and place. Growth regulation is discussed in Chapter 2, where it is pointed out that growth factors such as hormones play a very key role. Excess production of growth factors can have a tumor-inducing effect.

A different set of factors that inhibit, rather than promote, cell growth has recently been identified. Originally, cell growth inhibitors were called *chalones* to distinguish them from hormones. Recent examples of proteins with inhibitory properties are the interferons, tumor growth factor–β, and tumor necrosis factor (Feramisco et al., 1985). These and other factors are now being purified in various laboratories. Some of them, such as interferon and tumor necrosis factor, are already being tested clinically.

A tight interaction of growth promotors and growth inhibitors occurs in normal cells. Cancer clearly involves a disruption that is probably the consequence of genetic changes involving activation of growth promotors and inactivation of growth inhibitors. From the viewpoint of cancer therapy and prevention, we are particularly interested in identifying the inhibitors and adding them back to cells that have lost them through genetic changes.

Cell–Cell Interaction. Normal cells produce substances that suppress the growth of transformed cells. The inhibition of growth of polyoma virus transformed BHK cells (a Syrian hamster kidney cell line) by normal mouse fibroblasts was first described more than 20 years ago. In those experiments, a confluent monolayer of normal cells inhibited growth of the Py–BHK cells that were in contact with them.

Several investigators have proposed that cell–cell contact in these experiments is achieved through gap junctions, which are short connections between adjacent cells. This hypothesis has been tested directly in experiments combining growth measurements with direct examination of junctional communication by observing the transfer of fluorescent dyes from cell to cell (Mehta et al., 1986). Growth inhibition was correlated with the presence of junctional communication by varying the extent of communication. Drugs that alter the effectiveness of the junctions enhanced growth when junctional communication was blocked. Conversely, inhibition was maximized when junctions were open. Because only molecules below 1500 molecular weight can move across gap junctions, growth-inhibitory molecules in this small size range are being sought.

Tumor Angiogenesis Factor. Angiogenesis is the process by which a vascular network of blood vessels develops. Its importance in tumorigenesis has been recognized by physicians since the turn of the century. Tumors cannot grow above a tiny size and will regress unless they attract blood vessels for a sustaining blood supply. Tumor angiogenesis factor (TAF), a substance synthesized and secreted by tumor cells, is essential for their survival (Folkman, 1984). It follows that inhibition of TAF production would be a potential mode of tumor suppression.

A few cell constituents with TAF activity have been identified recently, and the genes that encode them have been cloned. They include basic fibroblast growth factor (bFGF) and angiogenin, a small protein of about 13,000 molecular weight. Once a gene is cloned, it becomes possible to study its regulation and try to devise methods to block its activity in tumor cells.

Suppression by Terminal Differentiation. Cells that have undergone terminal differentiation no longer divide and thus cannot give rise to tumors. One mode of suppression therefore consists of driving partially differentiated cells into the terminal state. The leukemias provide numerous examples of cells derived from specific stages in myeloid, lymphoid, or erythroid differentiation. Presumably, their differentiation pathways have been blocked by mutations or chromosome rearrangements. This would make the cells available for cancerous growth instead of terminal differentiation in the form of mature blood cells.

Similarly, in retinoblastoma one could suppose that the *Rb* gene codes for a protein that drives the retinal cells into final differentiation. Absence of this protein would permit further cell division and availability for oncogenesis. Thus, genes that promote differentiation can be considered a class of tumor-suppressor genes. Retinoids and other substances that induce differentiation in some model systems show promise in anticancer therapy (Sporn and Roberts, 1983).

Senescence. When normal human cells are grown in culture, they stop dividing after about 50–60 doublings of the population. This phenomenon is called senescence. The cells continue to metabolize, that is, to carry out internal chemistry. In this sense they are alive, but reproductively they are dead. No one has found a way to reverse the process experimentally. However, cancer cells do not senesce. One of the key genetic changes in cancer cells is immortalization. Cancer cells will grow indefinitely, like bacteria, if given suitable growing conditions.

Studies with human cells have revealed the importance of senescence in tumor suppression. Recent experiments showed that a retrovirus introduced into normal human cells could induce transient growth of small tumors when the infected cells were implanted into test mice (O'Brien et al., 1986). However, growth soon stopped as a result of senescence. Another sample of the same cells was grown in culture. It senesced at about the same number of population doublings as the sample inoculated into the mice.

This experiment provides a graphic example of a natural process that may prevent human cells from developing into life-threatening cancers. By providing a natural brake on cell growth, senescence acts as an inborn mechanism of tumor suppression.

The results suggest an explanation for some puzzling autopsy findings. Individuals who died of other diseases, as well as cancer, sometimes have small nodules of abnormal cells (termed premalignant) that have not developed into tumors. These few cells may have doubled enough times to have reached their reproductive senescence. Many cells die during tumor growth, because their chromosome rearrangements are usually lethal. Only a rare successful roll of the mitotic roulette wheel produces a viable tumor cell.

Senescence is a cellular process and should not be confused with longevity, which involves aging of the entire organism. Cellular senescence probably occurs in various tissues of the body, but the missing cells are replaced by others. Senescence is a form of terminal differentiation. From the point of view of cancer therapy, both terminal differentiation and loss of reproductive ability by senescence represent natural end-states of cells no longer capable of producing tumors.

In summary of this section, many mechanisms have developed to protect the long-lived human organism from cancer. As a result, the transformation of a normal cell to a tumor cell that grows to form a clinical cancer is a very rare event per cell. Unfortunately, it is all too frequent per person.

Cloning of Tumor-Suppressor Genes

Molecular identification of genes with tumor-suppressor activity is the most direct and powerful way of developing clinical applications of the tumor-suppressor concept. For this purpose, several laboratories are using new methods of isolating and cloning genes.

The underlying procedure is to introduce DNA from normal cells into tumor cells and then to select for those rare tumor cells that have been growth inhibited or altered as a result of incorporating a tumor-suppressor gene into their genome. Nongrowing cells can be selected under experimental conditions that kill dividing cells. For example, dividing cells will incorporate radioactive precursors of DNA such as tritiated thymidine or toxic analogs such as bromodeoxyuridine (BrdU) into DNA and subsequently be killed. Nondividing cells in the same mixed-cell population will survive because they are not synthesizing DNA. Another approach is to treat with drugs that kill tumor cells but not normal cells, on the assumption that the tumor-suppressed cells will be more resistant to such drugs than tumor cells. Both of these approaches are being used in various laboratories.

The next step is to isolate total cellular DNA from these suppressed cells and try to repeat the experiment by transferring this DNA into the same clone of tumor cells as was used initially. If the experiment is repeated successfully, then the DNA that was transferred into the tumor cells can be isolated and cloned. Once the tumor-suppressor gene has been cloned, it can be transferred into bacteria or other simple cells to express high concentrations of the encoded protein. This protein then becomes available to test as a possible clinical tool.

The cloning and characterization of genes that suppress tumor formation, whatever the mechanism, represents a large challenge and a great hope. This class of genes provides a new approach to cancer therapy and prevention based on the stimulation of the body's own protective mechanisms. The genetic aberrations that lead to cancer include the loss of activity of these genes. Can we restore the balance by supplying the organism with this lost activity? Gene cloning makes possible the manufacture of large quantities of either the gene or its gene product, a protein with tumor-suppressing activity. Thus, with the use of cloned normal gene products introduced at elevated concentrations, we may be able to restore the Yin–Yang balance of harmony and health.

References

Bodmer, W. F.; Bailey, C. J.; Bodmer, J.; Bussey, H. J. R.; Ellis, A.; Gorman, P.; Lucibello, F. C.; Murday, V. A.; Rider, S. H.; Scambler, P.; Sheer, D.; Solomon, E.; Spurr, N. K. "Localization of the Gene for Familial Adenomatous Polyposis on Chromosome 5." *Nature (London)* **1987,** *328,* 614–616.

Botstein, D.; White, R. L.; Skolnick, M.; Davis, R. W. "Construction of a Genetic Linkage Map in Man Using Restriction Fragment Length Polymorphisms." *Am. J. Hum. Genet.* **1980,** *32,* 314–331.

Boveri, T. *The Origin of Malignant Tumors;* Williams and Wilkins: Baltimore, MD, 1929 (Translated from the German volume of 1914).

Brauch, H.; Johnson, B.; Hovis, J.; Yano, T.; Gazdar, A.; Pettengill, P. S.; Graziano, S.; Sorenson, G. D.; Poiez, B. J.; Minna, J.; Linehan, M.; Zbar, B. "Molecular Analysis of the Short Arm of Chromosome 3 in Small-Cell and Non-Small-Cell Carcinoma of the Lung." *New Engl. J. Med.* **1987,** *317,* 1109–1113.

Cancer: Science and Society; Cairns, J., Ed.; W. H. Freeman and Co.: San Francisco, 1978; p 199.

Cavenee, W. K.; Hansen, M. F.; Nordenskjold, M.; Kock, E.; Squire, J. A.; Phillips, R. A.; Gallie, B. L. "Genetic Origin of Mutations Predisposing to Retinoblastoma." *Science (Washington, DC)* **1985,** *228,* 501–503.

Chan, J. Y. H.; Becker, F. F.; German, J.; Ray, J. H. "Altered DNA Ligase I Activity in Bloom's Syndrome Cells." *Nature (London)* **1987,** *325,* 357–359.

Chromosome Mutation and Neoplasia; German, J., Ed.; Alan R. Liss: New York, 1983; p 451.

Feramisco, J.; Ozanne, B.; Stiles, C. D. *Cancer Cells 3: Growth Factors and Transformation.* Cold Spring Harbor Laboratory: Cold Spring Harbor, 1985; p 450.

Folkman, J. "Angiogenesis." In *Biology of Endothelial Cells*; Jaffe, E. A., Ed.; Martinus Nijhoff: The Hague, The Netherlands, 1984; pp 412–420.

Housman, D. E.; Glaser, T.; Gerhard, D. S.; Jones, C.; Bruns, G. A. P.; Lewis, W. H. "Mapping of Human Chromosome 11: Organization of Genes Within the Wilms' Tumor Region of the Chromosome." *Cold Spring Harbor Symp. Quant. Biol.* **1986,** *51,* 837–841.

Knudson, A. G., Jr. "Hereditary Cancer, Oncogenes, and Antioncogenes." *Cancer Res.* **1985,** *45,*1437–1443.

Koufos, A.; Hansen, M. F.; Copeland, N. G.; Jenkins, N. A.; Lampkin, B. C.; Cavenee, W. K. "Loss of Heterozygosity in Three Embryonal Tumours Suggests a Common Pathogenetic Mechanism." *Nature (London)* **1985,** *316,* 330–334.

Mehta, P. P.; Bertram, J. S.; Loewenstein, W. R. "Growth Inhibition of Transformed Cells Correlates with Their Junctional Communication with Normal Cells." *Cell* **1986,** *44,* 187–196.

Murphree, A. L.; Benedict, W. F. "Retinoblastoma: Clues to Human Oncogenesis." *Science (Washington, DC),* **1984,** *223,* 1028–1033.

Nowell, P. C. "The Clonal Evolution of Tumor Cell Populations." *Science (Washington, DC)* **1976,** *194,* 23–28.

O'Brien, W.; Stenman, G.; Sager, R. "Suppression of Tumor Growth by Senescence in Virally Transformed Human Fibroblasts." *Proc. Natl. Acad. Sci. USA* **1986,** *83,* 8659–8663.

Rowley, J. D.; Ultmann, J. E. *Chromosomes and Cancer: From Molecules to Man*; Academic Press: New York, 1983; p 357.

Sager, R. "Genetic Suppression of Tumor Formation: A New Frontier in Cancer Research." *Cancer Res.* **1986,** *46,* 1573–1580.

Seizinger, B. R.; Martuza, R. L.; Gusella, J. F. "Loss of Genes on Chromosome 22 in Tumorigenesis of Human Acoustic Neuroma." *Nature (London)* **1986,** *322,* 644–647.

Solomon, E.; Voss, R.; Hall, V.; Bodmer, W. F.; Jass, J. R.; Jeffreys, A. J.; Lucibello, F. C.; Patel, I.; Rider, S. H. "Chromosome 5 Allele Loss in Human Colorectal Carcinomas." *Nature (London)* **1987,** *328,* 616–619.

Sparkes, R. S.; Murphree, A. L.; Lingua, R. W.; Sparkes, M. C.; Field, L. L.; Funderburk, S. J.; Benedict, W. F. "Gene for Hereditary Retinoblastoma Assigned to Human Chromosome 13 by Linkage to Esterase D." *Science (Washington, DC)* **1983,** *219,* 971–973.

Sporn, M. B.; Roberts, A. B. "Role of Retinoids in Differentiation and Carcinogenesis." *Cancer Res.* **1983,** *43,* 3034–3040.

Varmus, H. E. "The Molecular Genetics of Cellular Oncogenes." *Annu. Rev. Genet.* **1984,** *18,* 553–612.

Weissman, B. E.; Saxon, P. J.; Pasquale, S. R.; Jones, G. R.; Geiser, A. G.; Stanbridge, E. J. "Introduction of a Normal Human Chromosome 11 into a Wilms' Tumor Cell Line Controls its Tumorigenic Expression." *Science (Washington, DC)* **1987,** *236,* 175–180.

Willis, A. E.; Lindahl, T. "DNA Ligase I Deficiency in Bloom's Syndrome." *Nature (London)* **1987,** *325,* 355–357.

CHAPTER 6 Cancer as a Problem in Development

Frederick Meins, Jr.

Developmental biology is an important viewpoint from which to study cancer. This path has led to the concept that cancer cells can be coaxed back to a normal, or at least noncancerous, pattern of growth and specialization.

Normal development is an orderly and precisely regulated process. New body parts are formed with characteristic shapes. Cells become differentiated (i.e., specialized) to produce specific substances. They go through periods of active growth by enlargement and division. Some cells cease dividing at a certain point and never divide again over the life span of the organism; other cells continue to divide according to well-defined rules. Cancer profoundly alters this pattern of regulation.

Characteristics of Cancerous Growth

Growth Autonomy. When cells became cancerous—a process called neoplastic transformation or simply transformation—they either continue to divide or begin to divide when they should not. This capacity for "lawless" or inappropriate growth is called growth autonomy; it is an essential prop-

1420–4/88/0097$06.00/0

erty of all cancer cells. Autonomous growth does not necessarily mean the cells divide very rapidly. For example, if part of a rat's liver is surgically removed, normal cells in the remaining parts of the organ start dividing far more rapidly than certain types of cancerous liver cells. The critical difference is that the cancer cells divide continuously, whereas the normal cells stop dividing when the liver reaches a certain size.

Alterations in Cellular Heredity. The second essential property of cancer cells is that growth autonomy, once established, is inherited by individual cells. Toward the end of the last century it was recognized that cancers are transplantable. When small pieces of tumor tissues are implanted into the appropriate host organism, they develop into cancerous growths with properties very similar to those of the cancer from which they were derived. With more modern cloning methods, it has been possible to culture individual cancer cells in a nutrient medium that permits them to form a clone, which is a colony of cells descended from a single cell. The colonies often retain their cancerous character, demonstrating that growth autonomy can be transmitted to daughter cells.

Parallels in Cancerous and Normal Development

Neoplastic development—the formation of tumors from individual transformed cells—is a progressive multistep process that occurs over a period of years in human cancer. Like normal development, it appears to involve cell differentiation, the commitment of cells to specific fates, and cell migration. Differentiated cells in different organs produce specific chemical compounds. For example, crystallins are made only in lens cells of the eye, and albumin is made only in hepatocytes of the liver. Cancer cells are also capable of varying degrees of normal differentiation; certain tumors arising from the rat liver still produce albumin. In other cases, tumor cells produce differentiated products that are inappropriate to their original tissue. Bronchiogenic carcinomas and "oat cell" carcinomas of the human lung are able to produce trophic hormones normally elaborated exclusively by the pituitary gland. Generally, however, with increasing malignancy cancer cells tend to lose their capacity to produce cell-type specific substances.

How a normal cell develops depends on both its internal state and its surroundings. Signals from adjacent cells, from more distant parts of the body, and from the environment can trigger changes in the characteristics (the phenotype) of cells. This reactivity to signals, which is called competence, also changes during development. In the 1930s Spemann showed that when tissues of very young amphibian embryos from the region that would normally form an eye were grafted into the region that would become trunk, they formed trunk structures. The forming trunk appears to generate signals, called inducers, that can cause tissues to become trunk structures. The eye-forming tissues, on the other hand, are said to be competent to form trunk structures in response to the inducer.

Cellular competence and response to inductive signals also play a role in the development of cancers. The first step in carcinogenesis, initiation, is the conversion of a normal cell to a latent tumor cell. In a second step, promotion, the latent cells begin to divide and give rise to visible tumors. Thus, during initiation cells become competent to express the tumor phenotype, but only do so in response to some signal. Skin tumors can be induced in mice by chemical treatment. A single injection with urethane converts some skin cells into latent tumor cells. Without further treatment, however, no tumors actually form. This requires a second treatment with a promoter such as croton oil, which triggers the initiated skin cells to divide autonomously.

Cancer cells can change their competence. Certain mammalian tumors depend on hormones produced elsewhere in the body for their autonomous growth. For example, some breast cancers stop growing when deprived of a source of female sex hormones. Occasionally, cells in these tumors become hormone independent and thereafter transmit the hormone-independent state to their daughter cells. In general, states of competence and differentiation, like growth autonomy, can be inherited by cancer cells.

Cell Determination

Cellular inheritance of competence is not unique to cancer cells. During normal development, parts of organisms are committed to become specific structures in response to inducers. This process is called determination. Studies of the fruit fly, *Drosophila*, have shown that different states of determination can be inherited by individual cells. These cells exhibit their final differentiated phenotype only in response to specific hormonal signals. This "clonal" type of development seems to be particularly important in the formation of mammalian red and white blood cells. Stem cells, in response to specific growth- and differentiation-inducing factors, give rise to clones of daughter cells. These cells undergo sequential determination to yield the differentiated cell types that finally appear in the circulating blood. There is growing evidence that in human lymphoid malignancies (acute lymphoblastic and myeloid leukemias) transient cell types in the normal developmental pathway are somehow stabilized and continue to divide abnormally without further differentiation. In other words, the cells become inappropriately determined.

Genes, Gene Expression, and Cancer

What stabilizes cancer cells when they divide? All cell characteristics are specified by genes, which are the basic units of heredity. Thus, in principle, cancer could result from changes in the genes themselves or changes in the expression of genes. A wealth of evidence indicates that different types of highly specialized cells have the same complement of genes. There are a few exceptions, such as the mammalian red blood cell, which does not have a cell nucleus at maturity.

Specialized cells exhibit different properties because they express different genes. A striking demonstration of this important principle comes from studies of plants. It is possible to clone highly specialized plant cells and then to regenerate complete, fertile plants from the clones. This result demonstrates that the cells are totipotent; they contain all the genes necessary to form the complete organism (Figure 1).

Plant regeneration experiments have also shown that cells exhibiting different inheritable capacities for growth and production of growth factors remain totipotent. Thus, differences in gene expression can even account for determination of patterns of growth. This effect is called epigenetic (i.e., developmental) inheritance to distinguish it from permanent alterations in the genetic material of the cell.

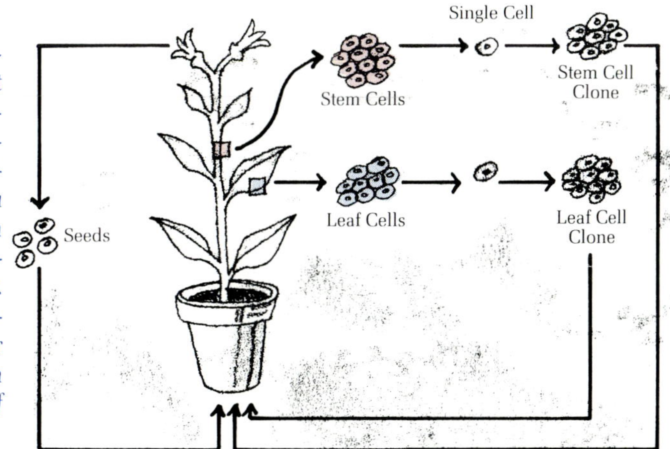

Figure 1. An experiment showing that highly specialized plant cells are totipotent and that cell-heritable determination is potentially reversible. Highly specialized leaf cells (blue) require a cell division factor for growth in culture; cells from near the surface of the stem (pink) do not. These two states of determination are transmitted to daughter cells. Fertile, normal plants can be regenerated from clones of both types of cells.

Students of tumor biology have tended to favor genetic rather than epigenetic hypotheses for cancer (Figure 2). There is good evidence that some types of cancer result from genetic changes. Two basic mechanisms are known. The first is mutation, in which DNA, the genetic material, is permanently changed by substitutions, duplications, losses, or rearrangements. The second mechanism is the introduction of foreign genes, usually derived from viruses, into the host cell during tumor initiation. These genes then become integrated into the genetic material and are transmitted to daughter cells when the cell divides.

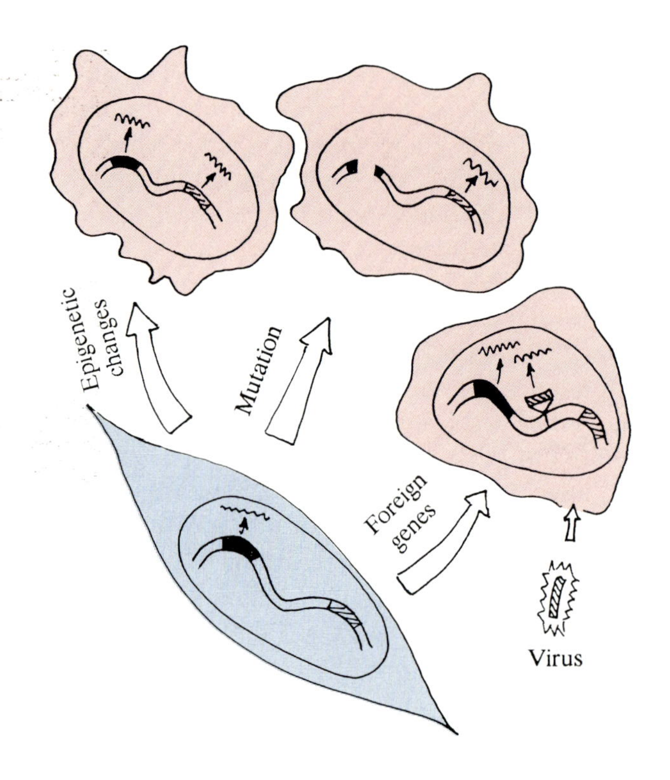

Figure 2. Genetic and epigenetic mechanisms for neoplastic transformation. Cells can be changed genetically by mutation, for example, by the loss of genetic material or by the addition of foreign genes from cancer-causing viruses. Cells can also undergo epigenetic changes similar to those that occur in normal development (see Figure 1). Here the genetic material of the cell is not permanently altered; instead genes that are silent in normal cells become expressed or genes expressed in normal cells become silent.

Even when cancer is caused by genetic changes, the actual formation of the cancer appears to be epigenetically controlled. Mutations that cause leukemia are known. Nevertheless, many types of cells bearing this mutation do not give rise to malignancies. Cancerous cells arise only from specific intermediate cells in the pathway leading to lymphocytes. Similar conclusions may be drawn for cells carrying cancer-causing genes (oncogenes) derived from viruses. Cellular expression of the malignant state depends on the differentiation of the cell.

This condition emphasizes further that neoplastic development is not an uncontrolled process. Cancer cells are susceptible to certain developmental and regulatory signals. Thus, in principle, it may be possible to change the state of differentiation of cancer cells and direct them back to a nonmalignant state.

Neuroblastoma. In some well-documented cases cancer cells revert to normal, either spontaneously or as part of their developmental program. A striking example is neuroblastoma, a highly malignant tumor of humans. This tumor appears to arise early in development and usually kills children in their first year of life. In up to 16% of the cases, a highly malignant, metastasizing cancer becomes a benign tumor. The neuroblastoma cells stop dividing and differentiate into what appear to be ganglion cells, which are normal constituents of the nervous system. This result clearly shows that certain cancer cells retain the potential for normal differentiation and growth regulation.

Skin Cell Carcinoma. Partial or even total reversion can also occur in squamous epidermoid carcinoma, a common type of skin cancer. In this case, the cancer cell behaves somewhat like a normal basal cell of the skin. When the cancer cell divides, one daughter cell differentiates to form a skin cell incapable of further division, while the other daughter cell retains its autonomous properties.

Chemically Induced Cancer of the Newt. In 1942 Needham proposed that it might be possible to "re-program" cancer cells for normal development by subjecting the cells to the inductive signals that generate new structures in parts of the body undergoing regeneration. Certain amphibia, such as the European newt, can readily regenerate their tails and limbs. Seilern-Aspang and Kratochwil induced highly malignant, metastasizing tumors at the base of the tail by treating newts with chemical carcinogens. Some of the cancer cells underwent differentiation and lost their malignant properties. This type of reversion was promoted by amputating the tail to stimulate regeneration near the tumor.

Crown Gall Disease. Studies of plant tumors have shown that cancer cells can be induced to develop normally, even when cancer results from genetic changes. When plants of many different species are wounded in the presence of a soil bacterium, *Agrobacterium tumefaciens*, large tumors form at the wound site. These crown gall tumors are composed of inheritably altered, autonomous-growing tumor cells that continue to grow abnormally in the absence of the bacterium. Neoplastic transformation is caused by the transfer of oncogenes from the bacteria to the host cell. These genes specify the inappropriate production by the host cell of auxins and cytokinins, which are substances needed for the growth of normal plant cells.

In the 1950s Braun showed that it was possible to regulate the malignancy of crown gall cells (Figure 3). Certain strains of the *Agrobacterium* induce a special type of tumor called a teratoma. Teratomas form highly abnormal organs and contain some cells that appear to differentiate normally. Cloned teratoma cells continue to exhibit their cancerous properties indefinitely when grown in culture. They also form teratomas when grafted onto the side of the stem of the host plant. In striking contrast, when the growing tip of a plant is removed and replaced with cloned teratoma cells, shoots arise from the grafted tissues. As they grow, the shoots gradually become normal in appearance. Eventually they flower and set seed. The seed can be germinated and develop into perfectly normal, fertile plants.

Figure 3. Crown gall disease. Teratoma cells (black) and teratoma tissues (black) derived from single teratoma cells contain oncogenes from the inducing bacterium and grow autonomously. When grafted at the tip of a host plant, the cloned teratoma tissue can form normal-appearing shoots (pink). The neoplastic state is suppressed in these shoots, even though the cells still contain the foreign oncogenes. Neoplastic growth resumes when tissues of the suppressed tumor shoots are returned to culture. During meiosis, the oncogenes are lost and seed from the suppressed tumor shoots develop into perfectly normal, nontumorous plants that do not contain the oncogenes (blue).

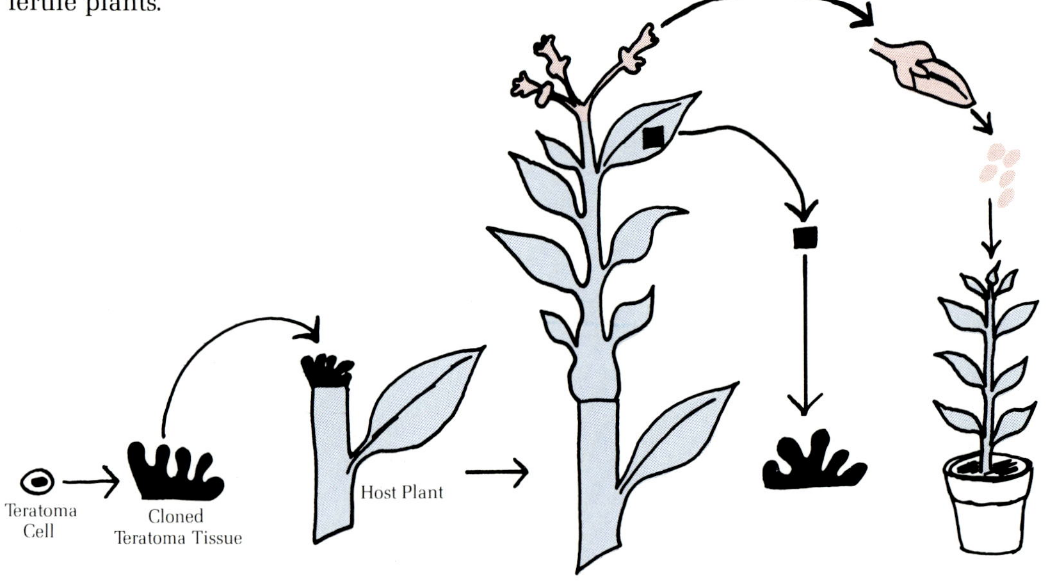

These experiments show that crown gall cancer cells, like normal differentiated cells, are totipotent. Progeny of teratoma cells can be directed to lose their malignant properties completely and to form plants. The plants grown from seed of teratoma cell origin no longer contain the oncogenes. Thus, in this case, tumor reversal is due to the loss of the foreign genes that originally caused the transformation. This loss occurs during meiosis, a particular stage in sexual reproduction.

Control of Tumors

Tumor Suppression. Crown gall teratoma cells can also be directed to develop normally without the loss of the foreign genes. This is called tumor suppression. Sensitive tests show that cells in the teratoma shoot prior to meiosis still contain the oncogenes, even though they differentiate and carry out specialized functions normally. Moreover, when individual specialized cells are removed from the shoot and placed in culture, they once again exhibit their cancerous properties. Thus, the teratoma shoots are actually composed of cells that are determined to be cancer cells but do not express their malignant character. Developmental influences acting in the shoots are strong enough to make cancer cells behave normally.

Mouse Teratocarcinoma. It may be argued that tumor reversal is a special property of certain amphibia and plants that have a high potential for regeneration and developmental "reprogramming". This is not the case. Examples of experimentally induced tumor reversal in mammals are well documented. The mouse teratocarcinoma is a highly malignant tumor that consists of a variety of differentiated cell types typical of normal tissues such as cartilage, bone, muscle, and nerves. These tumors also contain undifferentiated embryonal carcinoma cells, which are responsible for malignant growth. When individual embryonal carcinoma cells are placed in the abdominal cavity of mice, they give rise to malignant teratocarcinomas that eventually kill the animals (Figure 4).

Very different results were obtained when embryonal carcinoma cells were placed in early mouse embryos. In these experiments, the implanted cells carried specific biochemical markers so that it was possible to distinguish progeny of the cancer cells from cells of the host embryo. The embryos containing the cancer cells were placed in "incubator"-mother mice, where they developed into baby mice. The cancer-derived cells differentiated into many normal cell types, including liver, kidney, and thymus gland cells that are never found among the differentiated cells in the teratocarcinoma. Most of the adult mice containing the cancer-derived cells did not form cancers. Therefore, implanting cancer cells in early embryos in intimate contact with normal embryonic cells induces the differentiation and normalization of the cancer cell and its progeny.

Figure 4. Mouse teratocarcinomas contain a mixture of differentiated cells (colored) and potentially malignant embryonal carcinoma cells (black). When implanted in a mouse, the embryonal carcinoma cells form teratocarcinomas, which kill the animal. When implanted in an early mouse embryo, the embryonal carcinoma cells can normalize completely and form most of the cell types found in normal mice. Teratocarcinomas do not form in these mice.

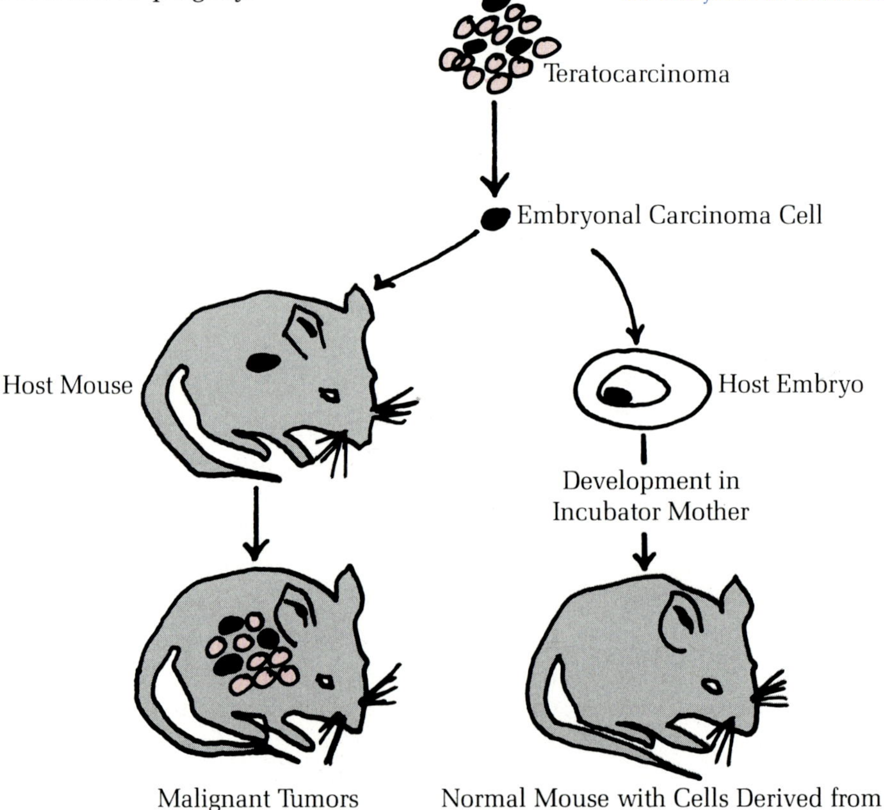

Normalization of Leukemia Cells. The examples cited so far provided good evidence that cancer cells can be induced to normalize by subjecting them to certain developmental influences. The crucial problem is to identify these influences. Are they specific chemical factors? Is contact with special types of cells required? Insight into this problem has come from studies of L. Sachs on the differentiation of human and myeloid leukemia cells grown in culture (Figure 5). A group of chemical substances that can regulate the differentiation of myeloid precursor cells has been identified. These substances are called monocyte and granulocyte inducers (MGI). When cultured human or mouse cells are treated with one of these inducers, MGI–2, they stop dividing, lose their malignant character, and differentiate into normal-appearing cells. Within a short time MGI–2 suppresses the population of leukemia cells.

This treatment also is effective in animals. Injection of MGI–2, or a chemical compound that induces MGI–2 production, inhibits the development of leukemia in mice inoculated with leukemia cells.

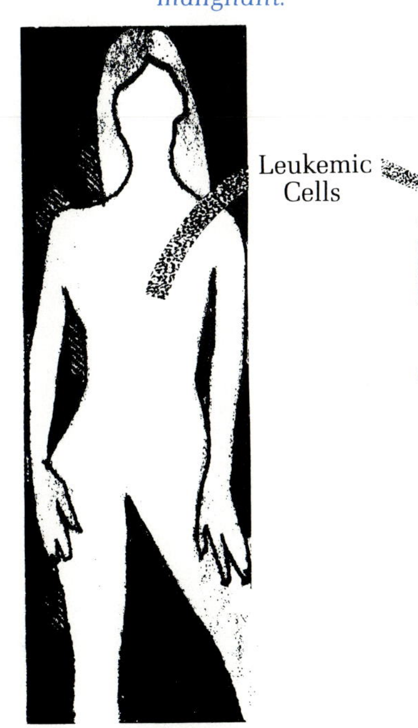

Figure 5. Leukemia cells can be cultured from leukemic mice or leukemia patients. The leukemia cells (pink) continue to grow in culture and retain their malignant character. Treating the cells with specific proteins needed for normal development of myeloid cells can induce the cells to become normal-appearing differentiated cells (blue) that do not divide and are no longer malignant.

Tumor Reversal and Cancer Therapy. The experiments just cited provide strong evidence that cancer is a problem in gene regulation. Even when the genetic material of the cell is altered by mutation or the introduction of foreign genes, expression of the cancerous state is ultimately regulated by epigenetic mechanisms like those operating in normal development. The course of tumor development is not fixed. The malignancy of some cancer cells can be regulated by factors that regulate normal cell differentiation.

At present we do not know whether tumor reversal is a general phenomenon. Very few of the chemical factors believed to regulate neoplastic development have been identified. Nevertheless, the experimental results are encouraging. They raise the hope that in the future it may be possible to treat cancer by inducing the normalization of malignant cells.

Suggested Reading

General

Braun, A. C. *The Biology of Cancer*; Addison-Wesley: Reading, MA, 1974.

Klein, G.; Klein, E. "Conditioned Tumorigenicity of Activated Oncogenes." *Cancer Res.* **1986,** *46,* 3211–3224.

Cell Commitment

Okada, T. S. "Commitment and Instability in Cell Differentiation." In *Current Topics in Developmental Biology*; A. A. Moscona; Monroy, A., Eds., Vol. 20, Academic Press: New York, 1986.

Meins, F., Jr., Foster, R. "Transdetermination of Plant Cells." *Differentiation (Berlin)* **1986,** *30,* 188–189.

Plant Tumors

Braun, A. C.; Wood, H. N. "Suppression of the Neoplastic State with the Acquisition of Specialized Functions in Cells, Tissues, and Organs of Crown Gall Teratoms of Tobacco." *Proc. Natl. Acad. Sci. USA* **1976,** *73,* 496–500.

Huff, M.; Turgeon, R. "Neoplastic Potential of Trichomes Isolated from Tobacco Crown Gall Teratomas" *Differentiation (Berlin)* **1981,** *19,* 93–96.

Meins, F., Jr. "Developmental Regulation of Tumor Autonomy in Plants." In *On the Nature of Cancer;* Müller, H., Ed.: Birkhäuser: Basel, 1985.

Neuroblastoma

Everson, T. C.; Cole, W. H. *Spontaneous Regression of Cancer*; Saunders: Philadelphia, 1966; p 500.

Squamous Cell Carcinoma

Pierce, G. B.; Wallace, C. "Differentiation of Malignant to Benign Cells." *Cancer Res.* **1971,** *31,* 127–134.

Mouse Teratocarcinoma

Mintz, B.; Fleishman, R. A. "Teratocarcinomas and other Neoplasms as Developmental Defects in Gene Expression." *Adv. Cancer Res.* **1981,** *34,* 211–278.

Leukemia

Greaves, M. F. "Differentiation-Linked Leukemogenesis in Lymphocytes" *Science (Washington, DC)* **1986,** *234,* 697–704.

Sachs, L. "The Reversibility of Neoplastic Transformation: Regulation of Clonal Growth and Differentiation in Hematopoiesis and Normalization of Myeloid Leukemia Cells." *Adv. Viral Oncol.* **1984,** *4,* 307–329.

CHAPTER 7 Cancer-Causing Chemicals

Elizabeth K. Weisburger

Chemists need to be able to recognize the possible adverse effects of chemicals. Early reports indicated that chemists have a higher than normal risk of developing cancers of the lymphatic system and pancreas. However, later studies concentrating on chemists, especially those who had been employed at companies with good health and safety programs, show lower death and cancer rates than those of the general population. Nevertheless, all chemicals should be treated with respect, particularly new ones that have not yet been tested. Unfortunately, it has become popular today for society to blame many of its ills on chemicals. We forget that we are composed of chemicals, that food is essentially chemical, that chemicals have facilitated the prevention and treatment of diseases, and that many comforts of modern life have a chemical basis.

Although cancer was known in ancient times, it is a current problem. Newspapers, radio, and television frequently inform us of cancer-causing agents in our environment. Air, water supplies, foods, almost everything we use, various industrial processes, wastes, and waste dumps all come under suspicion at one time or the other. Nevertheless, estimates that attribute human cancer to various causes indicate that about 4% may be due to occupational exposure to chemicals, 30% (in males) to tobacco use, 10% to sunlight, and 30–65% to lifestyle. About 10–15% is presumed due to unknown influences that may encompass effects of stress and other factors. Cancer rates and types around the world vary. They may change over a period of years, reflecting changes in the environment to which humans are exposed. Furthermore, since the death rates from such diseases as tuberculosis, smallpox, and pneumonia have declined, the general age of the population is increasing. Old age itself is one of the greatest risk factors for developing cancer.

Industrial Carcinogens

Associating exposure to chemicals with cancer incidence is hardly new. Such a relationship was first proposed by Percivall Pott in 1775, as a result of observing scrotal cancer in English chimney sweeps. Most sweeps entered the trade when very young, took few baths, and rarely changed their soot-covered clothing. Their extensive and prolonged exposure to soot tended to be followed by development of cancer in young adulthood. However, the actual carcinogenic substances in soot were not isolated and identified until over 150 years later (Chart I).

As European industrial development increased during the 19th century, workers in the shale oil and coal tar industries developed high rates of skin cancer. Dyestuff workers developed more bladder cancer than was usual. Within the past 20 years there has been an increased incidence of angiosarcoma of the liver, a very rare tumor, noted in people exposed to vinyl chloride. Lung cancers occurred rapidly in those working with chloromethyl methyl ether, usually contaminated with bis(chloromethyl) ether; this finding led to the discovery of another human carcinogen. Data for 24 substances, including some drugs, strongly suggest that they produce cancer in exposed humans. Some examples are the

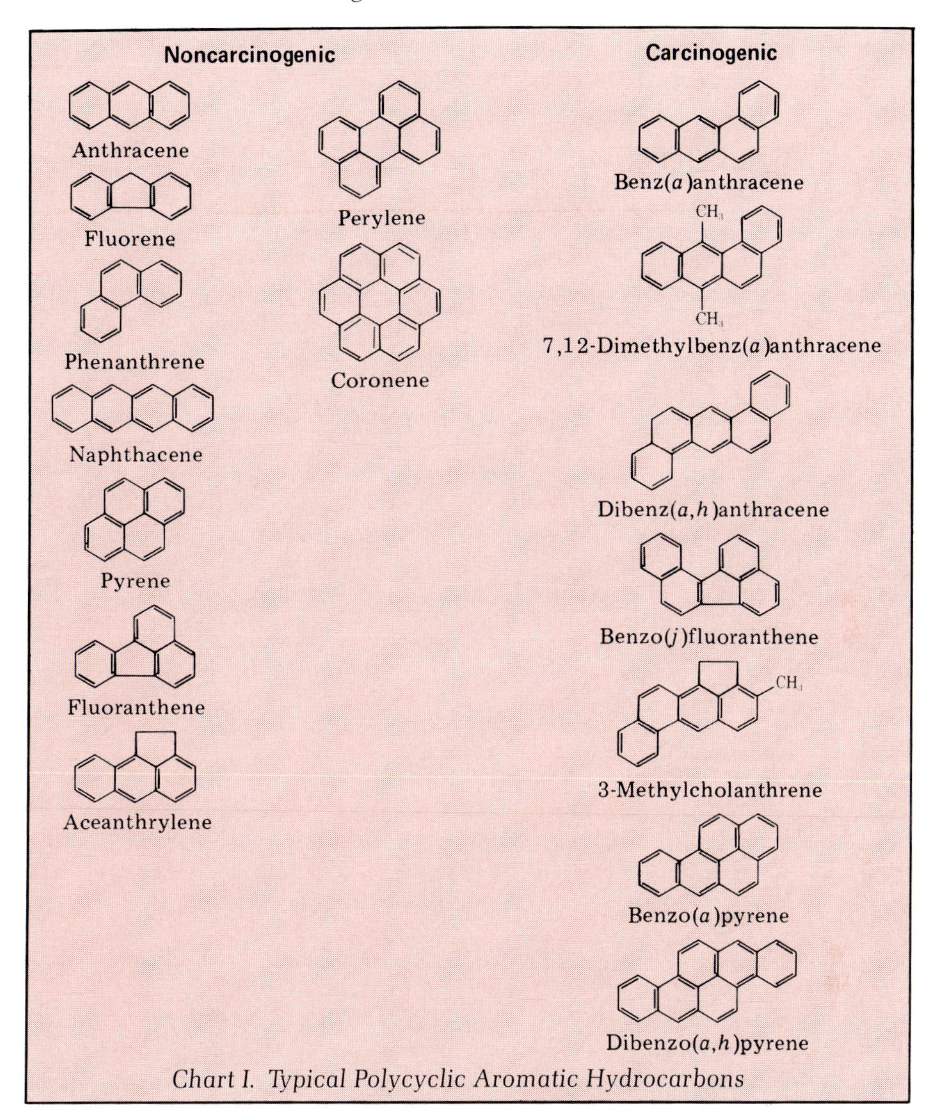

Chart I. Typical Polycyclic Aromatic Hydrocarbons

dyestuff intermediates benzidine and 2-naphthylamine (Chart II), bis(chloromethyl) ether, vinyl chloride, and benzene, among others. Research developments, beginning in 1915 when Japanese scientists painted rabbits with coal tar and thus induced tumors, have shown that these compounds can produce cancer in animals.

On the other hand, surveys of people who were exposed to some animal carcinogens have not shown evidence of any higher cancer rates than usual; saccharin is an example. The International Agency for Research on Cancer and the National Toxicology Program have made compilations of different substances, classifying them as associated with cancer in humans, probable carcinogens, and those for which there is no evidence that they are human carcinogens (see list).

Recognized Human Carcinogens

4-Aminobiphenyl
Analgesic mixtures containing phenacetin
Arsenic and certain arsenic compounds
Asbestos
Azathioprine
Benzene
Benzidine
N,N-Bis(2-chloroethyl)-2-naphthylamine
Bis(chloromethyl) ether and technical grade chloromethyl methyl ether
1,4-Butanediol dimethylsulfonate
Certain combined chemotherapy for lymphomas
Chlorambucil
Chromium and certain chromium compounds
Coke oven emissions
Conjugated estrogens
Diethylstilbestrol
Melphalan
Methoxsalen with ultraviolet A therapy
Mustard gas
2-Naphthylamine
Nickel refining
Thorium dioxide
Vinyl chloride

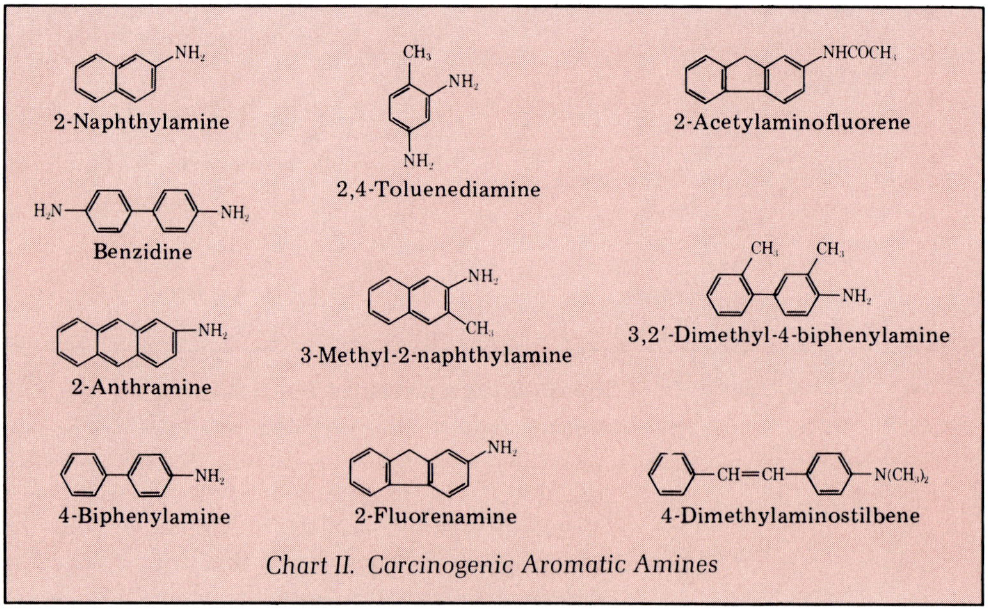

2-Naphthylamine
2,4-Toluenediamine
2-Acetylaminofluorene
Benzidine
3-Methyl-2-naphthylamine
3,2′-Dimethyl-4-biphenylamine
2-Anthramine
4-Biphenylamine
2-Fluorenamine
4-Dimethylaminostilbene

Chart II. Carcinogenic Aromatic Amines

Many compounds have been tested for carcinogenicity, some for research purposes, others as a prelude to possible widespread use (Chart III). Because of extensive programs during the past 20 years to test compounds for carcinogenic and other toxic properties, we know that many substances once considered innocuous have carcinogenic effects. An example is di(2-ethylhexyl) phthalate (DEHP), which produces liver tumors at high dose levels. DEHP is used as a plasticizer for poly(vinyl chloride) polymers. Studies on the mechanism for the effect of DEHP point toward involvement of the 2-ethylhexyl moiety, probably through inducing proliferation of peroxisomes in the liver. Similarly, a two-year test of unleaded gasoline by inhalation in rats and mice led to kidney cancers in male rats. Further investigation indicated that the isoparaffin, 2,2,4-trimethylpentane, was probably responsible.

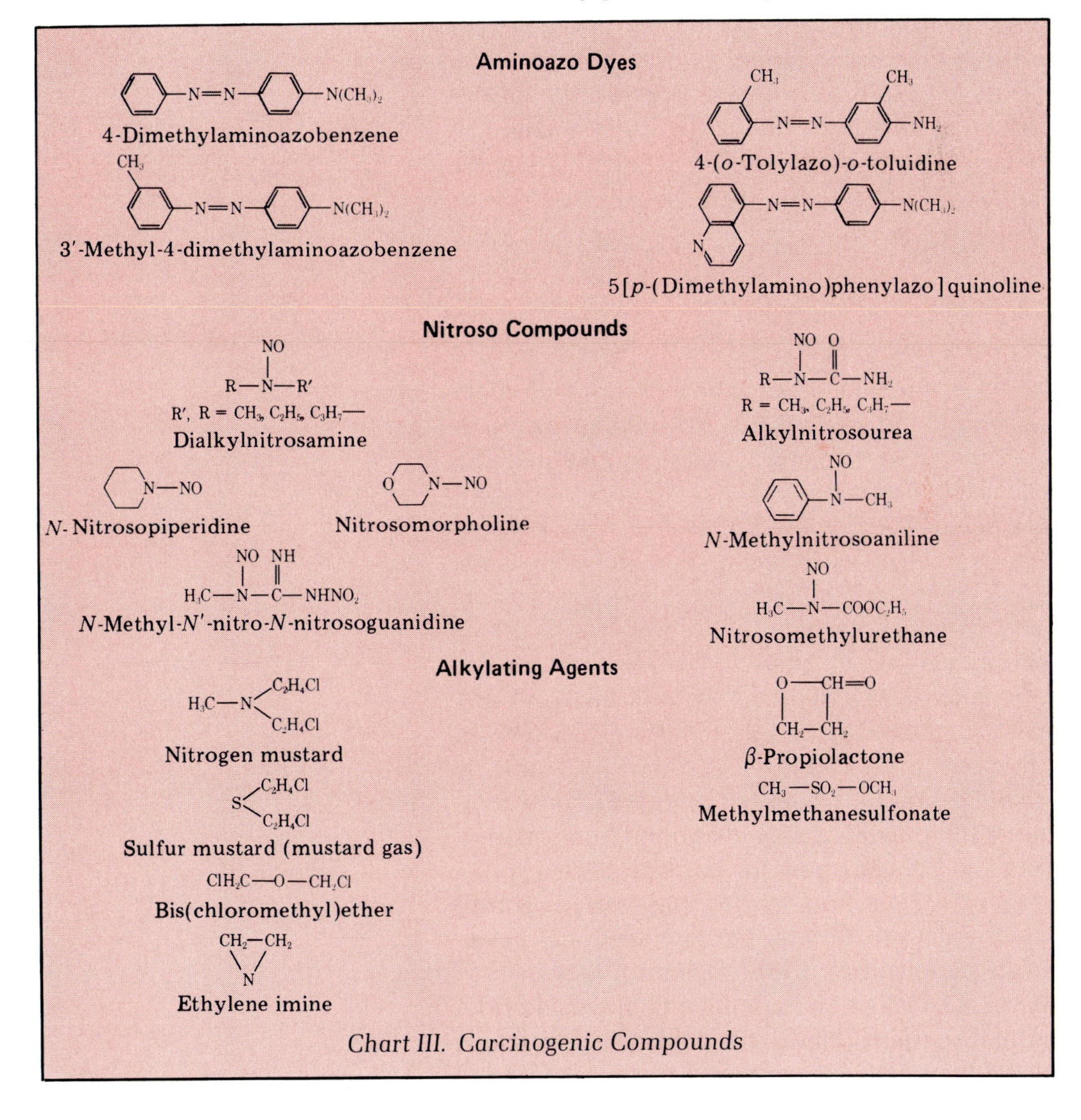

Chart III. Carcinogenic Compounds

The problem then arises whether the results at the high doses given the animals are relevant to humans who are exposed to much lower levels, for there is an assumption that substances that are carcinogenic in animals probably would be so in humans. As a consequence of these uncertainties, a new area, *risk assessment*, has arisen to deal with the possibility of developing cancer from exposure to very low levels of such compounds as DEHP. Those doing risk assessment may use slightly different assumptions, but all use some model to extrapolate from the high doses used in the bioassays or animal tests to the much lower levels to which people are exposed and to calculate the probable risk of developing cancer. A lifetime risk of 1 in 1,000,000 (10^{-6}) or less is presently considered societally acceptable. For comparison, the lifetime risk of a motor vehicle accident is 1 in 65, obviously much greater than the allowed risk of developing cancer.

Carcinogens in Food

Not all environmental carcinogens occur through industry. In the early 1960s an unknown disease killed thousands of young turkeys in England. The responsible factor was a toxin, aflatoxin, produced by a fungus, *Aspergillus flavus*, that contaminated the peanut meal fed to the turkeys. The same toxin, only this time in cottonseed meal, was the cause of many liver tumors in rainbow trout raised on a pelleted diet in fish hatcheries. Aflatoxin is carcinogenic to the livers of rats, ducks, trout, monkeys, and neonatal mice. Many other naturally produced substances are carcinogenic (Chart IV, Table I). Examples are the structurally related compounds, safrole from sassafras root, β-asarone from calamus roots, and estragole from the tarragon plant; cycasin and macrozamin from cycads and related species; allyl isothiocyanate from mustard seed and horseradish; pyrrolizidine alkaloids from plants such as ragworts, coltsfoot and comfrey; ptaquiloside from the bracken fern (*Pteridium aquilinum*) (eaten as the fiddleheads in Japan); and various hydrazine

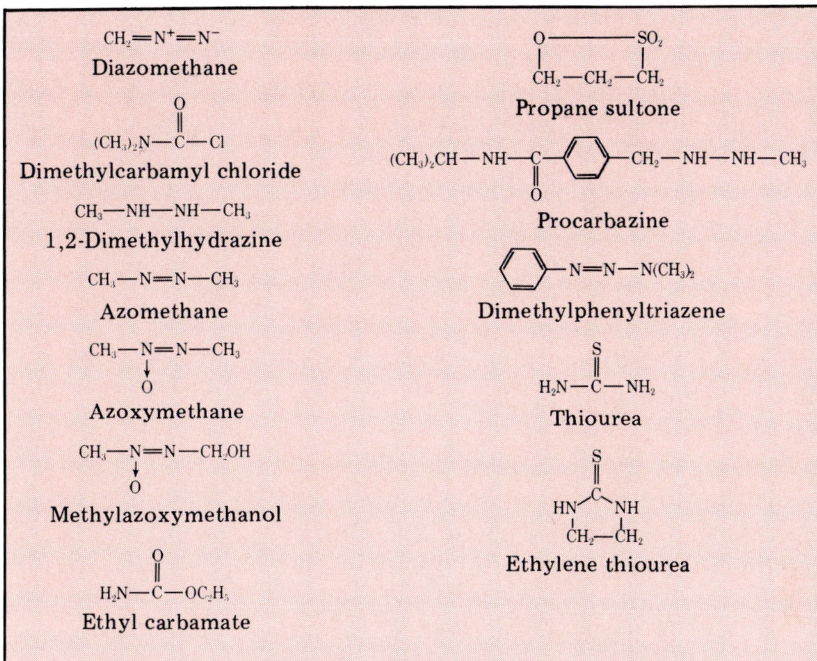

$CH_2{=}N^+{=}N^-$
Diazomethane

$(CH_3)_2N{-}C({=}O){-}Cl$
Dimethylcarbamyl chloride

$CH_3{-}NH{-}NH{-}CH_3$
1,2-Dimethylhydrazine

$CH_3{-}N{=}N{-}CH_3$
Azomethane

$CH_3{-}N(\rightarrow O){=}N{-}CH_3$
Azoxymethane

$CH_3{-}N(\rightarrow O){=}N{-}CH_2OH$
Methylazoxymethanol

$H_2N{-}C({=}O){-}OC_2H_5$
Ethyl carbamate

O—SO_2, $CH_2{-}CH_2{-}CH_2$ (ring)
Propane sultone

$(CH_3)_2CH{-}NH{-}C({=}O){-}C_6H_4{-}CH_2{-}NH{-}NH{-}CH_3$
Procarbazine

$C_6H_5{-}N{=}N{-}N(CH_3)_2$
Dimethylphenyltriazene

$H_2N{-}C({=}S){-}NH_2$
Thiourea

S, C, HN, NH, $CH_2{-}CH_2$ (ring)
Ethylene thiourea

Chart IV. Miscellaneous Carcinogenic Compounds

Table I. Origins of Some Carcinogenic Substances

Compound	*Origin*
Safrole	Sassfras root
B-Assarone	Calamus root
Estragole	Tarragon plant
Cycasin and macrozamin	Cycads and related species
Allyl isothiocyanate	Mustard seed and horseradish
Pyrrolizideine alkaloids	Ragworts, coltsfoot, and comfrey
Ptaquiloside	Brachen fern
Hydrazine derivatives	Mushrooms
Sterigmatocystin	Microorganisms
Ochratoxin	Microorganisms
Griseofulvin	Microorganisms
Chloroform	Marine creatures
Bromoform	Marine creatures
1,2-Dihaloethanes	Marine creatures
Benzene	Eggs, fruits, vegatables, cooked meats or fish, dairy products

derivatives made by common mushrooms, especially by the delicious false morel mushroom (*Gyromitra esculenta*). Other carcinogenic natural products include sterigmatocystin, ochratoxin, and griseofulvin, all products of microorganisms. Marine creatures produce chloroform, bromoform, and 1,2-dihaloethanes, all structures that merit attention. Benzene, although largely considered an industrial material, occurs naturally in eggs, fruits, vegetables, cooked meats or fish, and dairy products. It appears impossible to remove all carcinogens from the environment because trace amounts of so many are formed naturally.

More recent developments indicate that even the processing of foods or beverages can introduce carcinogens. Meats are usually cured with nitrite salts to inhibit the growth of the botulism organism. However, excess nitrite reacts with secondary amines and some of the free amino acids in the meat to produce N-nitroso compounds, some of which are carcinogenic in animals. The process of frying bacon can convert some of the noncarcinogenic nitrosamino acids thus formed to carcinogenic substances. Many compounds such as ascorbic acid, plant phenols, and some antioxidants can inhibit the nitrosation of secondary amines by competing with the amine for the nitrite. Vegetable foods generally contain nitrate; in turn, nitrate is reduced by the bacteria in the salivary plaque to nitrite, which can nitrosate secondary amines under the acidic conditions of the stomach. Experiments using the noncarcinogen nitrosoproline as a marker show that nitrosation does occur in humans. Browning of fresh meats and fish under usual frying conditions leads to formation of heterocyclic amines, which are pyrolysis products of amino acids, such as tryptophan. These substances are potent mutagens, and some are carcinogenic as well. However, the actual levels formed during cooking are extremely small—only 0.1 mg from almost 10 pounds of beef. Cooking by microwave apparently does not cause formation of these compounds.

In mice and infant rats, ethyl carbamate or urethane causes various types of tumors. Foods processed through fermentation, where ethanol is formed, contain very low but measurable levels of urethane. Ethanol probably reacts with carbamyl phosphate, present naturally in living organisms, to form urethane and phosphoric acid. As a consequence, such foods and beverages as olives, yogurt, bread, wine, beer, ale, and sake all contain urethane.

Within the past few years there has been a trend, especially among young men, to use snuff rather than to smoke cigarettes, which are themselves implicated as leading to lung and other cancers. The danger from this habit is quite high, as snuff contains nicotine-derived nitrosamines. In animal tests these compounds have shown considerable carcinogenic activity, especially for the esophagus and oral cavity. Furthermore, epidemiologic studies of people who have used snuff indicate a higher than usual rate of cancers of the oral cavity and esophagus.

Inorganic Carcinogens

Various inorganic substances such as certain salts of beryllium, cadmium, chromium, nickel, lead, and arsenic can lead to cancer in animals. Some industrial processes using chromates, arsenic, and nickel present a higher risk to workers. Nevertheless, the human body needs trivalent chromium to metabolize sugars, nickel is required by legumes and perhaps other plants, and arsenic may be a growth factor in rats.

The naturally occurring radioactive elements such as uranium, radium, and thorium and their daughters such as radon produce ionizing radiation that is also carcinogenic. Miners in European uranium ore regions and in the Colorado Plateau of the United States have higher cancer rates, particularly cancer of the lung, than normal; the effect is compounded by smoking. Furthermore, radon may accumulate in homes built over rock formations that are relatively rich in uranium. Living in homes built on the Colorado Plateau and over the Reading Prong, an underground rock formation extending through part of eastern Pennsylvania into New Jersey, may possibly increase the risk of cancer.

This risk can be reduced by proper ventilation in such homes.

Exposure to another naturally occurring inorganic material, asbestos, can also increase the risk of developing both lung cancers and mesothelioma, a rare type of tumor. The risk is increased many times by smoking. The size and shape of the fiber, not the composition, are critical factors. Another natural material, erionite, a zeolitic aluminosilicate that can occur in a fibrous form, is implicated as the cause of mesothelioma in certain Turkish villages, where it was often used as a building material. The concern regarding fibrous materials extends to various artifically made materials, such as mineral wool and fiberglass. Current manufacturing processes for these materials do not appear to be associated with excess risk of cancer.

Factors That Influence Carcinogenesis

An appreciable number of factors can influence the response of experimental animals to a specific carcinogen. Some of these are species, strain (of laboratory animal), sex, diet, immune status, metabolic rate, levels of certain enzymes, and age. For example the X/Gf strain mouse does not develop cancer if fed 2-acetylaminofluorene or ethyl carbamate, whereas other strains of mice show high incidences of tumors from such treatments. Diet is especially important, and there has been increased emphasis in this area lately. Feeding laboratory animals a restricted quantity of an adequate diet may decrease their response to some chemical carcinogens, as well as the usual spontaneous tumor incidence. On the other hand, feeding rats all they wanted of a diet deficient in choline and methionine increased the effect of a carcinogen. This deficient diet alone, without any carcinogen, led to a 40% incidence of liver tumors in the animals. Research on these modulating factors may increase means to inhibit or suppress the effects of chemical carcinogens.

How Carcinogens Act

Chemical carcinogens are divided into two broad classes: direct-acting and those requiring activation. The direct-acting carcinogens are usually electrophiles, that is, compounds seeking electrons. They react readily with nucleophiles (compounds rich in electrons) such as proteins, nucleic acids, and even water. Because of this reactivity, most direct-acting carcinogens are not likely to present a danger for the general population. However, those requiring activation are stable enough so that many people may be exposed environmentally or occupationally, with opportunity to ingest or absorb some of them before enzymes in liver, lung, or other organs convert them to their activated forms.

Not all electrophiles are equally active in leading to cancer. The shape of the atoms may determine whether the molecule fits into some cellular receptor. Solubility may determine whether it passes through cell membranes to attack target molecules within the cell. Reactivity determines whether the electrophile reaches its target before it reacts with cellular water. For example, nitrogen mustard, which is very reactive and has a short half-life, is not so potent a carcinogen as uracil mustard, which is less reactive and shaped more like the body's molecules.

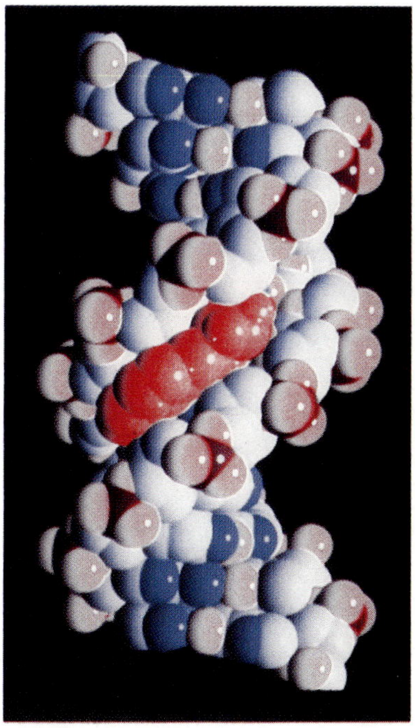

Much importance has been attributed to the attachment of activated carcinogens to the nucleic acids of the cell—the DNA, which contains the genetic information, and the RNA, which is involved in protein synthesis. There have been many attempts to correlate the degree of attachment at a particular position in the DNA with the relative potency of various carcinogens, all based on the concept that DNA interaction may trigger cancer. New developments have complicated these ideas. Some carcinogens react with multiple or all possible sites in the nucleic acids. Other animal carcinogens, such as carbon tetrachloride and chloroform, appear to have minimal or no interaction with nucleic acids. In general, the direct-acting compounds attach to the 7-position of guanine (Figure 1). Aromatic amines are more likely to form adducts

Figure 1. Computer graphic representation of a benzo[a]pyrene DNA adduct in which the orange spheres (benzo[a]pyrene plane) are inserted into the groove of the DNA.

at the C-8 and O-6 of guanine. Nitrosoethylurea reacts at all possible positions. Aflatoxin B_1 forms several adducts besides the one at the N-7 of guanine, but only a few have been identified positively. The major adduct from the prototype polycyclic aromatic hydrocarbon, benzo[*a*]pyrene, is from the 10-position of the hydrocarbon to the amino group in the 2-position of guanine. As for vinyl chloride, experiments with rat liver microsomes led to separation of three adducts, one at the 7-position of guanine, namely 7-*N*-(2-oxoethyl)-guanine. Two adducts were isolated in which the vinyl chloride had formed new rings, namely 1,N^6-ethenodeoxyadenosine and 3,N^4-ethenodeoxycytidine. However, only 7-N-(2-oxoethyl)guanine could be isolated from rats exposed to radiolabeled vinyl chloride, a result illustrating that experiments in vitro cannot always be correlated with those in vivo. Although the nucleotide base guanine is the main target for adduct formation, the role of minor adducts that may be more persistent, such as O^4-ethyldeoxythymidine from diethylnitrosamine, must be identified.

Even if chemical products formed with nucleic acids lead to distortion of the DNA structures and thus interfere in their normal biological activity, enzyme systems can repair the DNA. If this were not so, ordinary exposure to sunlight or natural background levels of radiation could lead to appreciable DNA damage. The repair enzymes cut out the damaged section of DNA containing the carcinogenic product and then insert a new section or "patch" in place of the damaged piece. Damage may be repaired more rapidly at certain regions than at others; this is especially true at the 7-position of guanine. If the system is overwhelmed by continued exposure to a carcinogen for an extended time, loss of repair capabilities results.

It has become possible to detect some of the DNA-carcinogen adducts in the tissues of exposed animals or persons by means of relatively specific and very sensitive immunoassay procedures. Femtomole (10^{-15}) levels of the benzopyrene-DNA adduct are readily detected in lung tissue from deceased smokers who had lung cancer, but not in lung tissue from patients who died of other causes. Such procedures may aid in determining whether there has been occupational exposure to specific carcinogens. There has been renewed interest as well in alterations in the proteins and polypeptides of the cell during stages of carcinogenesis. Earlier research showed that carcinogens bound to the methionine or histidine in proteins, but there was little effort devoted to studying the consequence of this binding. Recent investigations have led to identification of two protein adducts of 4-aminobiphenyl. One involved attachment to cysteine after oxidative metabolism of the amine; the other, a tryptophan adduct, must have involved acetylation of 4-aminobiphenyl, which is recognized as a carcinogen. These studies may lead to a method for identifying 4-aminobiphenyl adducts in the blood of smokers. This aromatic amine occurs in cigarette smoke, and smokers have a higher rate of bladder cancer than usual.

Metabolism

Carcinogens are metabolized by the same enzyme systems that metabolize other foreign compounds (xenobiotics). The purpose of most metabolic systems is to render xenobiotics water-soluble, so they may be conjugated and excreted from the body more readily. Therefore, the major reactions involve detoxification, and only a small fraction is generally metabolized to an active form. Examples of activation pathways are given in Chart V. The enzyme system generally involved is the cytochrome P–450 monooxygenase family, which has a diversity of isoenzymes. In turn, specific isoenzymes can be induced or enhanced by certain compounds and inhibited by still others. Other oxidative schemes may also be involved, such as prostaglandin H synthase. Additional enzymes, such as acetyltransferase, sulfotransferase, epoxide hydrolase, and glutathione transferase, are involved in certain activation and deactivation steps.

Experiments in laboratory animals indicate several stages in the process of carcinogenesis. These experiments have shown that the effects of

Alkylating agents

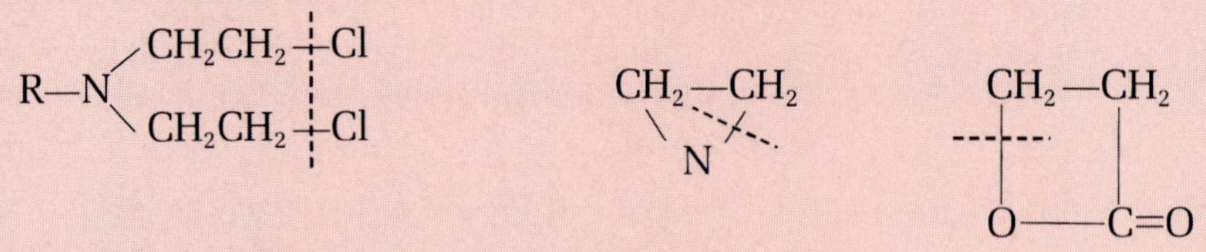

$R{-}N(CH_2CH_2{-}Cl)_2$ $\quad$ $CH_2{-}CH_2$ / N $\quad$ $CH_2{-}CH_2$ / O—C=O

Nitrosamines and related compounds

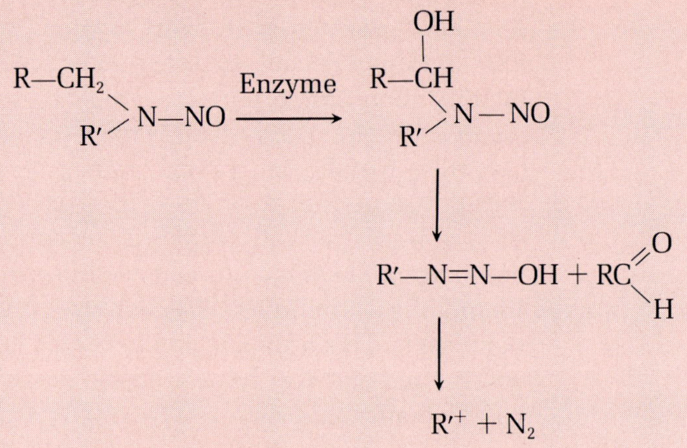

$R{-}CH_2(R')N{-}NO \xrightarrow{\text{Enzyme}} R{-}CH(OH)(R')N{-}NO$

$\downarrow$

$R'{-}N{=}N{-}OH + RC(=O)H$

$\downarrow$

$R'^{+} + N_2$

Polycyclic aromatic hydrocarbons

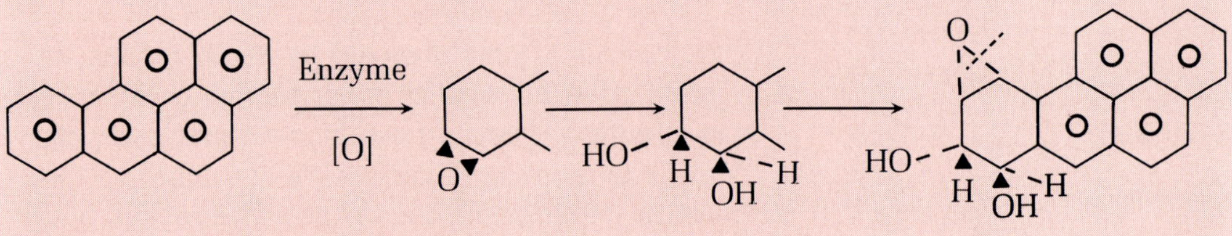

Enzyme [O] $\longrightarrow$ O $\longrightarrow$ HO, H, OH, H $\longrightarrow$ O, HO, H, OH, H

Aromatic amines and amino azo dyes

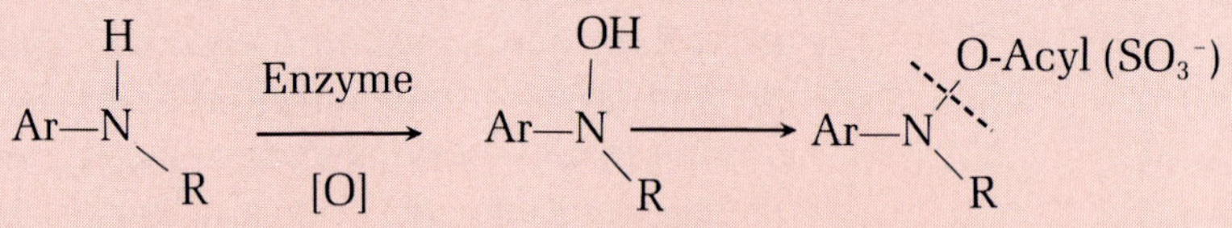

$Ar{-}N(H){-}R \xrightarrow[\text{[O]}]{\text{Enzyme}} Ar{-}N(OH){-}R \longrightarrow Ar{-}N(\text{O-Acyl }(SO_3^-)){-}R$

Safrole

O—CH_2, O, $CH_2{-}CH{=}CH_2$ $\longrightarrow$ O—CH_2, O, $C(OH){-}CH{=}CH_2$ (Acyl)

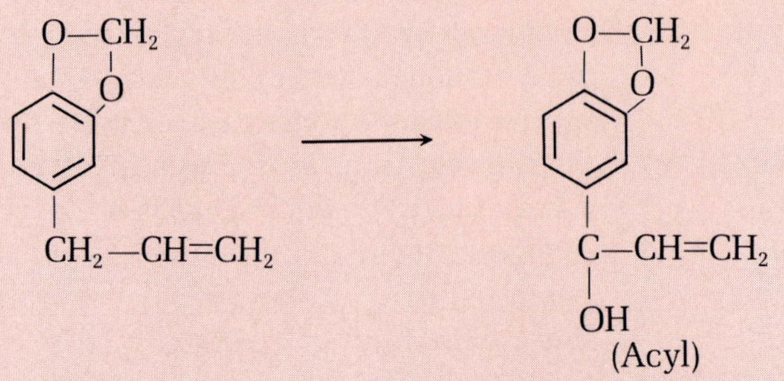

Chart V. Activation Mechanisms of Chemical Carcinogens

carcinogens can be overcome, increased, or modified by other compounds applied at some of these stages. Substances that enhance or increase the effect when applied after the carcinogen are called promoters. Most prominent among these are phorbol esters (terpenelike compounds), originally isolated from an ornamental tropical plant. However, certain compounds from bacteria and marine blue-green algae also are promoters, as are some naturally occurring amino acids. Cocarcinogens enhance the effect when applied with the carcinogen; an example is catechol, which has no action on its own but increases that of benzo[*a*]pyrene.

The opposite type of action, decreasing the effect, has been noted for many compounds. Retinoic acid derivatives or precursors, ellagic acid, ascorbic acid, vitamin E, selenium (as selenium salts), and indoles from Brussels sprouts are only a few of the many compounds tested for this purpose, some in experimental animals and some in clinical situations.

The mechanism of action for a few of these compounds has been investigated (Figure 2). Retinoic acid derivatives reverse the formation of excess keratin in cells. Selenium is a component of the glutathione peroxidase system, an enzyme that detoxifies many xenobiotics, and indoles are enzyme inducers that also increase the levels of detoxifying enzymes.

Such experiments increase the expectation that it is possible to influence the course of cancer. Human cancer can apparently be decreased by proper diet, which includes less fat, less red meat, less pastries or similar items, and more complex carbohydrates and fiber-rich whole grain products. More fish, green and yellow vegetables, and fruits should be eaten. Elimination of smoking and the use of other tobacco products can lower cancer rates appreciably. Avoidance of excessive exposure to sunlight and excessive use of alcohol are other positive measures, as is adoption of a moderate but varied lifestyle. With sufficient dedication and determination in the population, this type of cancer prevention may be possible.

KINDS OF REACTIONS

Alkylating capability. Agents add alkyl groups such as CH_3, into a molecule. Examples are nitrogen mustards, diazomethane, activated epoxides or aziridines, and chloromethyl ethers.

Alcylating capability. Agents add acyl group, such as:

$$-\underset{\underset{O}{\|}}{C}CH_3$$

into a molecule. Examples include β-propiolactone, propane sultone, and dimethyl carbamyl chloride.

KINDS OF COMPOUNDS

Aromatic amines. For these carcinogens, changing the amino group's position in a molecule can alter carcinogenicity. For example, pure 1-naphthylamine is not carcinogenic but 2-naphthylamine is. Addition of other groups also changes carcinogenicity. Adding a methyl group to the 3-position of 2-naphthylamine increases its carcinogenicity. Groups which increase polarity or water solubility often decrease carcinogenicity. Substitution by fluorine or chlorine might enhance it, but the larger halogens, iodine, or bromine, might reduce it.

Amino azo dyes. Same as for aromatic amines.

Nitrosodialkylamines, nitrosoamides, and nitrosoureas. For $R-\overset{\overset{NO}{|}}{N}-R$ *nitrosamines, compounds are usually active carcinogens if R = R′-alkyl. If R and R′ differ, the esophagus is a likely target organ.*

Most nitrosoamides and -ureas are carcinogenic. An exception is N-nitroso-N-methyl-p-toluenesulfonamide. Nitrosoureas are potent, often at a single dose, and can lead to unusual tumors, such as of the nervous system. Although N-nitroso heterocycles (compounds containing rings that have oxygen, nitrogen, or sulfur atoms) such as nitrosomorpholine and nitrosopiperazine are active, the N-nitroso amino acids are not carcinogens.

Polynuclear aromatic hydrocarbons. Such compounds with more than 120 A^2 in area, are immediately suspect for carcinogenicity. Substitution of methyl groups alters carcinogenicity. For example, benz(a)anthracene is weak but 7,12-dimethylbenz(a)anthracene is potent. Such compounds, when linear, are often noncarcinogenic.

Alkylthioureas. Alkylthioureas often lead to thyroid tumors in animals. Diarylthioureas are apparently noncarcinogenic.

Azomethane derivatives and precursors. Compounds such as azomethane, azoxymethane, methylazoxymethanol, 1,2-dimethylhydrazine, and aryldialkyl triazenes are all carcinogenic. Such activity depends on their decomposition, to methyl, CH_3^+; nitrogen, N_2; and formaldehyde, HCHO. The methyl, CH_3^+, might alkylate molecules in cell (see alkylation, above).

Miscellaneous. Numerous other organic compounds are carcinogenic. Examples include acetamide, 3-aminotriazole, aramite, carbon tetrachloride, chloroform, 1,2-dibromoethane, diethylstilbestrol, dioxane, ethionine, ethyl carbamate, and hexamethylphosphoramide (Chart IV).

Figure 2. Carcinogenic reactions and chemical classes.

References and Suggested Reading

Ames, B. N. "Dietary Carcinogens and Anticarcinogens." *Science (Washington, DC)* **1983,** *221,* 1256–1264.

Artvinii, M.; Baris, Y. I. "Malignant Mesotheliomas in a Small Village in the Anatolian Region of Turkey: An Epidemiologic Study." *J. Natl Cancer Inst.* **1979,** *63,* 17–22.

Blair, A.; Stewart, P.; O'Berg, M.; Gaffey, W.; Walrath, J.; Ward, J.; Bales, R.; Kaplan, S.; Cubit, D. "Mortality Among Industrial Workers Exposed to Formaldehyde." *J. Natl. Cancer Inst.* **1986,** *76,* 1071–1084.

Dipple, A.; Michejda, C. J.; Weisburger, E. K. "Metabolism of Chemical Carcinogens." *Pharmacol. Therap.* **1985,** *27,* 265–296.

Doll, R.; Peto, R. "The Causes of Cancer: Quantitative Estimates of Avoidable Risks of Cancer in the United States Today." *J. Natl. Cancer Inst.* **1981,** *66,* 1191–1308.

Eskew, D. L.; Welch, R. M.; Cary, E. E. "Nickel: An Essential Micronutrient for Legumes and Possibly All Higher Plants." *Science (Washington, DC)* **1983,** *222,* 621–623.

Gram, T. E.; Okine, L. K.: Gram, R. A. "The Metabolism of Xenobiotics by Certain Extrahepatic Organs and Its Relation to Toxicity." *Ann. Rev. Pharmacol. Toxicol.* **1986,** *26,*259–291.

Halder, C. A.; Holdsworth, C. E.; Cockrell, B. Y.; Piccirillo, V. J. "Hydrocarbon Nephropathy in Male Rats: Identification of the Nephrotoxic Components of Unleaded Gasoline." *Toxicol. Ind. Health* **1985,** *1,* 67–87.

Hecht, S. S.; Castonguay, A.; Rivenson, A.; Mu, B.; Hoffmann, D. "Tobacco Specific Nitrosamines: Carcinogenicity, Metabolism, and Possible Role in Human Cancer." *J. Environ. Sci. Health, Environmental Carcinogenesis Reviews* **1983,** *1,* 1–54.

Hirono, I. "Carcinogenic Principles Isolated from Bracken Fern." *CRC Crit. Rev. Toxicol.* **1986,** *17,* 1–22.

International Agency for Research on Cancer. IARC Monographs on the Evaluation of the Carcinogenic Risk of Chemicals to Humans. Chemicals, Industrial Processes and Industries Associated with Cancer in Humans, IARC Mongraphs, Vols. 1 to 19, IARC Monographs Supplement 4, IARC, Lyon, France, 1982.

Laib, R. J.; Gwinner, L. M.; Bolt, H. M. "DNA Alkylation by Vinyl Chloride Metabolites: Etheno Derivatives or 7-Alkylation of Guanine?" *Chem.-Biol. Interact.* **1981,** *37,* 217–231.

Loeb, L. A.; Ernster, V. L.; Warner, K. E.; Abbotts, J.; Laszio, J. "Smoking and Lung Cancer: An Overview." *Cancer Res.* **1984,** *44,* 5940–5958.

Mikol, Y. B.; Hoover, K. L.; Cresia, D.; Poirier, L. A. "Hepatocarcinogenesis in Rats Fed Methyl-Deficient, Amino Acid-Defined Diets." *Carcinogenesis* **1983,** *4,* 1619–1629.

National Research Council, Committee on Nitrite and Alternative Curing Agents in Food: The Health Effects of Nitrate, Nitrite, and N-Nitroso Compounds. National Academy: Washington, DC, 1981.

National Research Council, Committee on Diet, Nutrition and Cancer: Diet, Nutrition, and Cancer. National Academy: Washington, DC, 1982.

National Research Council, Committee on the Institutional Means for Assessment of Risks to Public Health: Risk Assessment in the Federal Government: Managing the Process. National Academy: Washington, DC, 1983.

National Toxicology Program. Fourth Annual Report on Carcinogens. U.S. Department of Health and Human Services, 1985.

Nebert, D. W.; Eisen, H. J.; Negishi, M.; Lang, M. A.; Hjelmeland, L. M. "Genetic Mechanisms Controlling the Induction of Polysubstrate Monooxygenase (P–450) Activities." *Ann. Rev. Pharmacol. Toxicol.* **1981,** *21,* 431–462.

Poirier, M. C.; Nakayama, J.; Perera, F. P.; Weinstein, I. B.; Yuspa, S. H. "Identification of Carcinogen–DNA Adducts by Immunoassays." In *Application of Biological Markers to Carcinogen Testing*; Milman, H. A.; Sell, S., Eds.; Plenum: New York, 1983; pp 427–440.

Ross, M. H.; Lustbader, E. D.; Bras, G. "Dietary Practices of Early Life and Spontaneous Tumors of the Rat." *Nutr. Cancer* **1982,** *3,* 150–167.

Chemical Carcinogens; Searle, C. E., Ed.; 2nd Ed. Vol. 1 and 2; ACS Monograph No. 182; American Chemical Society: Washington, DC, 1984.

Skipper, P. L.; Bryant, M. S.; Tannenbaum, S. R.; Groopman, J. D. "Analytical Methods for Assessing Exposure to 4-Aminobiphenyl Based on Protein Adduct Formation." *J. Occup. Med.* **1986,** *28,* 643–646.

Sugimura, T. "Studies on Environmental Chemical Carcinogenesis in Japan." *Science (Washington, DC),* **1986,** *233,* 312–318.

Swenberg, J. A.; Dyroff, M. C.; Bedell, M. A.; Popp, J. A.; Huh, N.; Kirstein, U.; Rajewsky, M. F. "O^4-Ethyldeoxythymidine, but not O^6-Ethyldeoxyguanosine, Accumulates in Hepatocyte DNA of Rats Exposed Continuously to Diethylnitrosamine." *Proc. Natl. Acad. Sci. USA* **1984,** *81,* 1692–1695.

Toth, B. "Synthetic and Naturally Occurring Hydrazines and Cancer." *J. Environ. Sci. Health, Environmental Carcinogenesis Reviews* **1984,** *C2,* 51–102.

Turnbull, D.; Rodricks, J. V. "Assessment of Possible Carcinogenic Risk to Humans Resulting from Exposure to Di(2-ethylhexyl)phthalate (DEHP)." *J. Am. Coll. Toxicol.* **1985,** *4,* 111–145.

Weisburger, E. K. "Chemical Carcinogenesis and Its Relevance for the General Population." In *Cancer Causing Chemicals;* Sax, N. I., Ed.; Van Nostrand Reinhold: New York, 1981; pp 3–13.

Weisburger, E. K. "Cancer-Causing Chemicals." In *Cancer: The Outlaw Cell;* LaFond, R. E., Ed.; American Chemical Society: Washington, DC, 1978; pp 73–85.

Wise, R. W.; Zenser, T. V.; Kadlubar, F. F.; Davis, B. B. "Metabolic Activation of Carcinogenic Aromatic Amines by Dog Bladder and Kidney Prostaglandin H Synthase." *Cancer Res.* **1984,** *44,* 1893–1897.

CHAPTER 8 Cancer-Causing Radiation

Robert L. Ullrich

Radiation can cause cancer. That simple fact was known by the early 1900s (Storer, 1975). Further, radiation can induce cancer in almost any tissue in animals and humans. But the cancer-causing dose may vary by 20-fold for different tissues in animals. Because such variation is also seen in people, the minimum dose that causes human cancer is not known. Thus, the crucial question becomes what factors, including amount of exposure, trigger cancer.

Radiation is divided into two types: ionizing and nonionizing. Of the two, ionizing radiation involves higher energies. As electrons are ejected from molecules, charged particles called ion pairs are formed. They are short-lived and often break down to form highly reactive free radicals, which are molecular fragments containing unpaired electrons. Nonionizing radiation, which involves ultraviolet light, microwaves, and radio waves, causes molecular excitations such as vibrations and electron movement, but produces no ions. Though ultraviolet light causes skin cancer, ionizing radiation is, by far, the more potent carcinogen (Storer, 1975; Ullrich, 1982).

1420–4/88/0131$06.00/0

Ionizing Radiation

Ionizing radiation falls into two classes: particulate (involving β particles, neutrons, and α particles) and electromagnetic (involving X-rays and γ rays). Each kind produces ionizations, but the way in which the ionizations are produced and the patterns may be different.

For example, because α particles are heavy, they produce ions along short paths and in dense clusters. By contrast, β particles (which are essentially electrons), X-rays, and γ rays leave longer paths of damage with wider spaces between sets of ion pairs (Figure 1). Since they are not charged particles, neutrons produce no ionizations directly. Instead they produce ionizations in living matter mainly by interaction with hydrogen nuclei (because living matter is composed mainly of water) and the production of protons. These charged protons in turn produce ionizations.

Because the sets of ion pairs that are produced can be spaced either close together or far apart, we refer to radiation's average energy released per unit distance as *linear energy transfer* (LET). The LET depends on the energy and charge of a particle. The greater the charge and lower the velocity, the greater is the LET. High-LET radiation (such as neutrons or α particles) produces dense sets of ion pairs, and low-LET radiation (such as β particles, X-rays, or γ rays) leaves sets of ion pairs that are spread out (Figure 2) (Storer, 1975).

High-LET radiation is more potent than low-LET radiation because of its more disruptive effects on biologically important molecules such as proteins, ribonucleic acid (RNA), and deoxyribonucleic acid (DNA). Radiation damage to DNA, the storehouse of genetic information, is considered most important.

Damage can occur in either or both strands of the DNA double helix. Many cells contain enzymes that can repair most damage that involves only one strand by taking advantage of the complementary character of the two strands (Figure 3) (Cleaver, 1986). Double-strand damage is more difficult to repair and more likely to be repaired incorrectly.

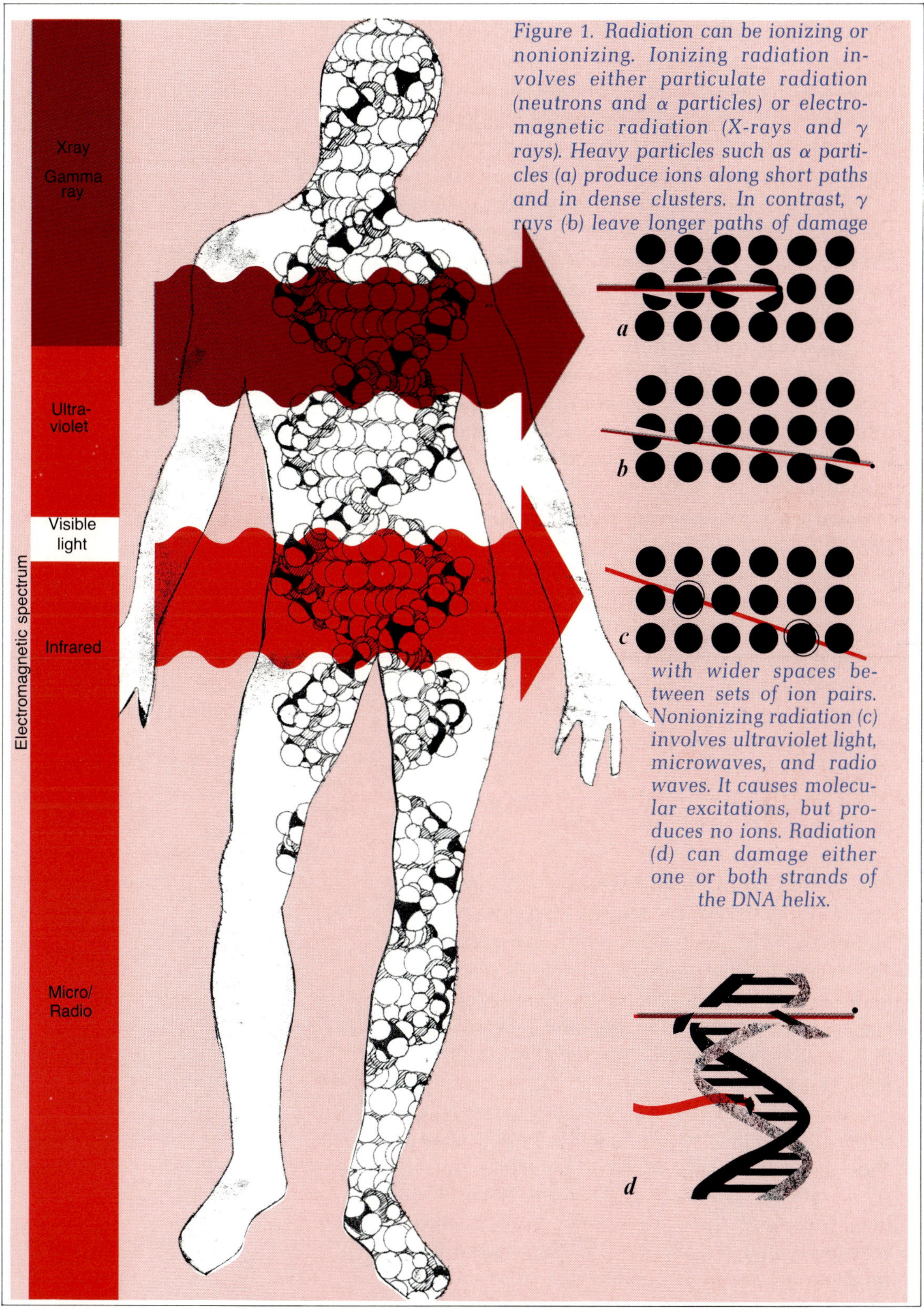

Figure 1. Radiation can be ionizing or nonionizing. Ionizing radiation involves either particulate radiation (neutrons and α particles) or electromagnetic radiation (X-rays and γ rays). Heavy particles such as α particles (a) produce ions along short paths and in dense clusters. In contrast, γ rays (b) leave longer paths of damage with wider spaces between sets of ion pairs. Nonionizing radiation (c) involves ultraviolet light, microwaves, and radio waves. It causes molecular excitations, but produces no ions. Radiation (d) can damage either one or both strands of the DNA helix.

High-LET radiation from even a single particle is a probable instigator for such damage, which depends not so much on dose as on the density of the ion track. This is because the greater the density of the track, the more likely it is that damage will occur in nearby places in the two DNA strands.

The goal of research on the cancer-causing effects of radiation is two-fold: to understand the risks to humans associated with exposure and to understand the mechanisms of cancer development. Information on the cancer-causing effects of radiation has come directly from exposed human populations and from experimental studies. These two sources provide different types of information about cancer development following radiation exposure. Studies of human populations provide direct information about the kinds of tumors produced by radiation, when tumors develop, what factors may influence susceptibility, and in some cases the relationship between radiation dose and cancer development. Experimental studies provide more precise quantitative information on the relationship between cancer development and radiation dose for a wide range of radiation types, allow the study of modifying factors under carefully controlled conditions, and allow the study of mechanisms.

Human Studies

Cancer is the only long-term cause of death that has been increased by moderate-to-low doses of radiation (Kohn and Fry, 1984). The purpose of studies on human populations is to develop a reliable quantitative understanding of the risks of cancer development following exposure to radiation. Of principal concern are the risks from doses of 10 rad or less.

Information about radiation and cancer comes from sources such as populations receiving exposures at work (for example, radiologists and uranium miners), for medical purposes (therapy and diagnosis), and from the atomic bomb blasts at Hiroshima and Nagasaki. All of these populations received doses higher than the dose range of

The gray (Gy), which is the unit of radiation absorbed dose, is equal to 1 Joule absorbed per kilogram of body tissue. The estimated average dose per person each year in the United States is 0.182 cGy. Natural background radiation accounts for 0.102 cGy, and most of the rest comes from medical exposure.

Generally, significantly detectable increases in cancer require from 50 cGy for sensitive tissues such as blood-forming cells to 10 Gy for resistant tissues such as skin. Below these doses, effects are blurred by other factors such as normal incidence of disease.

Relative biological effectiveness, RBE, is a way of comparing different kinds of radiation to a standard, usually γ rays. Comparisons are based on radiation needed to produce particular biological effects, such as mutations, cell killing, or tumors, and thus vary considerably.

For example, 1 Gy of γ rays might produce the same number of tumors in mice as would 0.2 Gy of fission neutrons; thus neutrons have an RBE of 5. Moreover, neutron RBE increases with decreasing dose, probably because the effects of low-LET neutron radiation are less at low doses.

Figure 2. Measuring radiation.

Figure 3. Three major known enzymatic repair processes. (a) Photoreactivation is effective in repairing ultraviolet light damage. A single enzyme binds to the damaged segment. After absorbing light, it splits the damaged covalent bonds, allowing them to fall back into place. (b) Excision repair involves a series of enzymatic steps to remove damaged sections of DNA and to code for replacements, by using complementary DNA strands as templates. (c) Postreplication repair fills existing gaps in strands newly synthesized from damaged DNA templates. (d) Double-strand damage is more difficult to repair. When damage is extensive, repair often is not possible.

interest, so risks cannot be measured directly but must be inferred from what is observed at higher doses. Such a calculation requires the development of a mathematical description of the dose–response relationship.

Several problems make this relationship difficult to model. Obviously, these human exposures are not laboratory experiments, and doses must usually be estimated. When patients are exposed for diagnosis or therapy their doses are known, but complicating illnesses may cloud interpretation. It is also difficult to identify a proper match among unexposed people, considering factors such as economic status, medical history, personal habits, occupation, and diet. And, of course, few studies involve large populations.

Because the increased numbers of cancers caused by radiation are relatively small compared to the number normally found in a population, large populations are required to detect these increases with statistical certainty. For example, the largest population exposed to radiation consists of the survivors of the atomic bombings in Japan. A major study of this population includes 54,580 irradiated subjects. So far 3832 cancers have been found. The cancer rate seen in the control population indicates that 3525 cancers would normally be found in the same-sized nonirradiated population. This statistic means that probably only 307 cancers were caused by radiation—an increase over the control incidence of about 8%. In a small population such an increase would be difficult to detect.

All these problems generate considerable uncertainty about what mathematical model best describes the dose–response curve. For low-LET radiation, both linear and linear quadratic models seem appropriate (Figure 4). A linear dose–response model is proportional to dose and has the mathematical form: incidence $= D$. A linear quadratic dose–response model is more S-shaped and has the mathematical form: incidence $= \alpha D + \beta D^2$. This S-shaped curve consists of a shallow linear response at low doses, followed by a rapidly rising response at intermediate doses. Higher doses bring a plateau in the response or a reduced response. This decline is attributed to competition between cancer induction and cell killing, because dead cells cannot lead to cancer (Upton, 1977). Although both models fit the dose, the predicted effects at low doses (10 rads or less) differ considerably. Most scientists favor the linear quadratic model, which best fits the pattern most commonly seen in animal studies and is supported by theories of how radiation interacts with matter (BEIR, 1980). For high-LET radiation there is even less information, but current experimental studies suggest a linear response. This linear assumption is also supported by theoretical considerations.

The group of atomic bomb survivors, the world's largest and most thoroughly studied irradiated population, is the principal source of information. The kinds of cancers that are increased in the atomic bomb survivors because of radiation exposure include leukemia; multiple myeloma; and cancers of the lung, breast, stomach, colon, and urinary tract (Kohn and Fry, 1984). A period of time elapses after radiation exposure before cancers are seen. This time is referred to as the latent period. The latent period for leukemia development is short, and so leukemias are the first cancer seen after exposure. The minimum latent period for leukemia is about two years in the atomic bomb survivors. Leukemias have not continued to develop after an initial postexposure wave. Instead, the number of leukemias was higher for the first 20 years, and since then no new leukemias have been seen. For other cancers the latent period is 10 years or longer. Cancer incidence following irradiation appears to be higher for the lifetime of the individual.

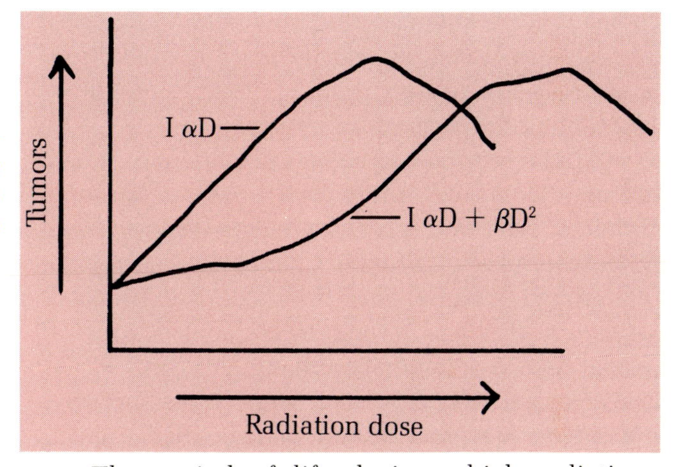

Figure 4. Typical linear and linear quadratic dose-response curves. The decrease in tumors at higher doses is thought to be due to cell killing.

The period of life during which radiation-induced cancers other than leukemias occur is the same time period during which naturally occurring tumors appear (Kohn and Fry, 1984). For example, breast cancer usually develops in women who are 35–40 years of age or older. An increased number of breast cancers in irradiated women is not seen until these women reach about 35–40 years of age. The reasons for this age dependency are not known, but it indicates that cancer development is influenced by many factors, including the person's physiological state. Whatever the reason, exposure early in life results in a longer latent period. Because of this longer latent period, it is just now being appreciated that in many cases persons exposed at younger ages tend to develop more cancers following a dose of radiation than do older persons receiving the same dose. A good example of this is radiation-induced cancer of the breast in females.

Experimental Studies

The cancer-causing effects of radiation have been studied with animals and, more recently, with cells. One of the goals of these experiments is to determine how radiation dose affects the frequency of cancer. Another goal is to understand how the cancer-causing effects of radiation are affected by the rate at which the dose is delivered and by the LET of the radiation. The third goal of experimental studies is to try to understand how radiation causes cancer. Animal studies usually involve irradiation of animals and observation of them throughout their lifetime for tumor development.

Cellular studies use either cells that are grown and maintained in culture (*cell lines*) or, in some instances, cells that have just been removed from animals (*primary cultures*). Changes in cellular in vitro growth potential have been shown to be good indicators of when a cell has changed from a normal cell to a tumor-producing cell (Hall et al., 1986). Normal cells grow as a single layer and stop growing when a tissue culture plate is filled. Cancer cells pile up and continue to grow. When a plate of cells is stained with dye, these piled-up colonies stain darker in dishes than normal cells, and so the cancer cells can be detected. Because the number of cells at risk is known, this change, called malignant transformation, can be quantified (Figure 5).

When animals are exposed to low-LET radiation the dose–response curves for tumor development vary, depending upon the kind of tumor induced. In the majority of instances the response resembles a linear quadratic curve. When the dose of low-LET radiation is delivered over a long period of time rather than rapidly, its effects are generally reduced. This is true for many radiation effects, including cell killing, cancer development in animals, and malignant transformation in cells (Ullrich, 1982; Kohn and Fry, 1984). Apparently, if given time, animals and cells can partly recover from radiation effects. This recovery might be caused by cells repairing themselves. Differences in the effects of the dose rate on the physiological state of the animal or the cellular environment could also affect whether a potential tumor cell develops into a tumor.

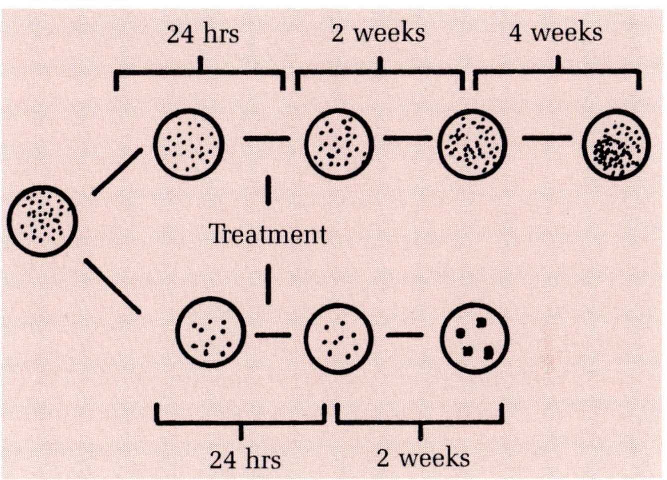

Figure 5. Illustration of an in vitro transformation experiment using a cell line. Growing stock cells are made into single-cell suspensions and plated on new Petri dishes. The top panel illustrates the procedure when a group of cells is plated to determine transformation. At the same time another group of cells (bottom panel) is plated to determine how many cells are killed by the treatment, so that transformation rates per viable cells can be determined. For the transformation experiment, enough cells are plated to result in about 400 viable cells remaining after treatment. For the cell survival experiment, further cells are needed (about 40 viable cells). Twenty-four hours after plating, the cells are irradiated. About two weeks later the dishes in the cell survival study can be stained, the cell colonies counted, and cell survival calculated. The transformation experiment takes four more weeks (a total of six weeks). Transformed cells can be detected through staining; colonies in which the cells are piled up on each other appear very dark.

Densely ionizing high-LET radiation is more effective than low-LET radiation for cancer induction and the induction of malignant transformation in vitro. The dose–response curve in both cases is linear rather than S-shaped. It has also been found in animal and cellular studies that reducing the dose rate of high-LET radiation does not reduce its cancer-causing effects (Fry, 1981).

As with humans, such factors as age, physiological state, and sex of the animal can all influence the cancer-causing effects of radiation. Young animals usually are more sensitive to cancer induction with radiation than are older animals. Certain tumor types such as ovary and breast tumors are expected to be sex-related, but other kinds of tumors also seem to be influenced by sex. For example, in certain strains of mice, males tend to be more sensitive than females to the development of myeloid leukemia (Ullrich, 1980).

Malignant transformation of a cell (at least its expression or detection) can be markedly influenced by its neighboring cells. Recently it has been shown that normal cells appear to suppress the growth of malignantly transformed cells in culture

(Farber, 1982). Normal cells can also inhibit the ability of transformed cells to grow into tumors in animals.

Mechanisms

The primary effects of radiation that are responsible for causing cancer are thought to involve effects on DNA. These effects could be associated with specific radiation-induced genetic mutations or perhaps with the activation of oncogenes. Oncogene activation is receiving much attention at present. Radiation is very effective at producing breaks in chromosomes (Wolff and Carrano, 1986). Such breaks often lead to the movement of pieces of DNA from one chromosome to another. These movements are called *translocations* (Figure 6) and are thought to be responsible for the activation of certain cellular oncogenes. In fact, specific translocations have been associated with cancer development in several specific instances.

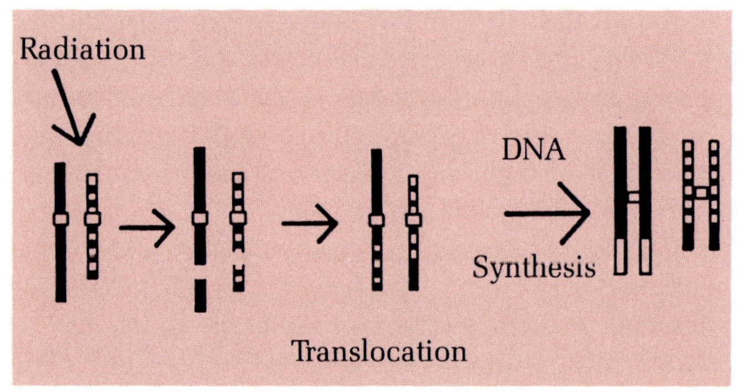

Figure 6. Chromosome translocations can be induced by radiation that produces breaks in two separate chromosomes, followed by rejoining in an inappropriate manner. Following DNA synthesis, the translocation is replicated and passed on. Such changes are usually not lethal, although other kinds of chromosome changes probably kill cells.

Although the initial radiation effects are at the level of DNA, other factors also play very important roles in determining whether these potentially cancer-causing cells actually develop into tumors. The study of these modifying factors, such as age, hormones, and cellular interactions, should provide basic information on cancer development. In addition, the apparently important role of such secondary factors offers the hope that methods might be devised to reduce the cancer-causing effects of radiation.

Acknowledgment

Research sponsored by the Office of Health and Environmental Research, U.S. Department of Energy, under contract DE-AC05-84OR21400 with the Martin Marietta Energy Systems, Inc., and by Grant CA 43322 from the National Cancer Institute.

References

Beir *The Effects on Populations of Exposure to Low Levels of Ionizing Radiation*, Report of the Committee on the Biological Effects of Ionizing Radiations, National Academy of Sciences, Washington, D.C., 1980.

Cleaver, J. E. "DNA Damage and Repair." In *Radiation Carcinogenesis*; Upton, A. C.; Albert, R. E.; Burns, F. J.; Shore, R. E., Eds.; Elsevier: New York, 1986; p 43.

Farber, E. "Chemical Carcinogenesis - A Biologic Perspective." *Am. J. Pathology* **1982,** *106,* 271.

Fry, R. J. M. "Experimental Radiation Carcinogenesis: What Have We Learned." *Radiat. Res.* **1981,** *87,* 224.

Hall, E. J.; Phil, D.; Hei, T. K. "Oncogenic Transformation of Cells in Culture: Pragmatic Comparisons of Oncogenicity, Cellular and Molecular Mechanisms." *Int. J. Radiation Oncology Biol. Phys.* **1986,** *12,* 1909.

Kohn, H. I.; Fry, R. J. M. "Medical Progress." *N. Engl. J. Med.* **1984,** *310,* 504.

Storer, J. B. "Radiation Carcinogenesis." In *Cancer*; Becker, E. F., Ed.; Plenum: New York, 1975; p 453.

Ullrich, R. L. "Radiation Carcinogenesis." In *Radiotracers in Biology and Medicine, Vol. VI, Radiation Biology*; Pizzarello, D. J., Ed.; CRC: Boca Raton, FL, 1982; p 111.

Ullrich, R. L. "Carcinogenesis In Mice After Low Dose and Low Dose Rate Irradiation." In *Radiation Biology in Cancer Research*; Meyn, R. E.; Withers, H. R., Eds.; Raven: New York, 1980; p 309.

Upton, A. C. "Radiobiological Effects of Low Doses: Implications for Radiological Protection." *Radiat. Res.* **1977,** *71,* 51.

Wolff, S.; Carrano, A. V. "Radiation-Induced Chromosome Aberrations and Cancer." In *Radiation Carcinogenesis*; Upton, A. C.; Albert, R. E.; Burns, F. J.; Shore, R. E., Eds.; Elsevier: New York, 1986; p 57.

CHAPTER 9 Cancer and the Immune Response

John L. Fahey and Brian C.-S. Liu

Immunology has made enormous progress during the past 65 years, bringing a significant improvement in the understanding of human disease and the relief of human suffering. Once-catastrophic infectious diseases can now be prevented by immunological procedures such as vaccination and antiserum injection, which direct the body's natural defense system to destroy invading organisms. It was only natural that these successful approaches against infectious diseases should be tried against cancer. If the immune system could be activated to destroy cancer cells in the patient's body without harming normal cells, it would be a potent means of cancer control. Unfortunately, malignant cells that develop in humans do not elicit the same types of immune response as infectious organisms or organ transplants. Because tumors grow in spite of low levels of immune response, the problems involved in tumor immunology are complex.

Organ transplant studies have made researchers increasingly aware of the highly specific and complex antigenic makeup of the individual. The biological reasons for this degree of individual differences are not clear. Perhaps, on the cellular level, it is for elimination of native cells that have become malignant.

1420–4/88/0143$06.00/0

The Immune System

The immune system is a powerful and complex defense mechanism that helps the body protect itself from foreign tissue and invading microbes such as viruses, bacteria, and parasites. It distinguishes such threats from normal tissue by recognizing antigens, or foreign molecules, and responding according to the nature of the antigen. Two types of immune cells, lymphocytes and macrophages, are involved in the body's counterattack. Macrophages are known as scavenger cells. They recognize, engulf, and dispose of damaged cells, cellular debris, and foreign invaders. The macrophages that line body cavities provide a first line of defense against microbial infections and parasitic infestations. They also participate in surveillance against foreign invaders and cancer.

Lymphocytes, a type of white blood cells, play a major role in the immune response. Dispersed throughout the body, these cells migrate through tissues, circulate in blood and lymph, and tend to accumulate in the spleen and lymph nodes. The two classes of lymphocytes, T-cells and B-cells, have distinct functions and can act independently. More commonly, however, they collaborate in their attack (Figure 1).

Both B-cells and T-cells originate in the bone marrow. During their development, however, T-cells migrate to the thymus, where they mature. When released after maturation, they can destroy foreign substances either by launching a direct attack or by activating other components of the immune system. Most T-cells cannot recognize free antigen circulating in the blood or lymph. They can respond to antigen on a cell surface only when the foreign substance is linked with one of the host cell's own proteins. Such molecules, coded by segments of DNA, make up the major histocompatibility complex (MHC). For an immune response to occur, the antigen receptor on the surface of a T-cell must simultaneously recognize the foreign antigen and the MHC protein. On recognizing an antigen, these T-cells multiply and differentiate into a variety of subclasses. Some are "killer" cells that specifically destroy any target cell with the appropriate surface antigen by disrupting the target cells' membranes and lysing them. Others produce substances that cause local inflammation. Still other T-cells control the responses of various antigen-triggered lymphocytes by promoting the maturation of antigen-stimulated B-cells and T-cells or by helping in the self-regulation of the immune response. Direct attack on invading foreign organisms by whole cells is called cell-mediated immunity.

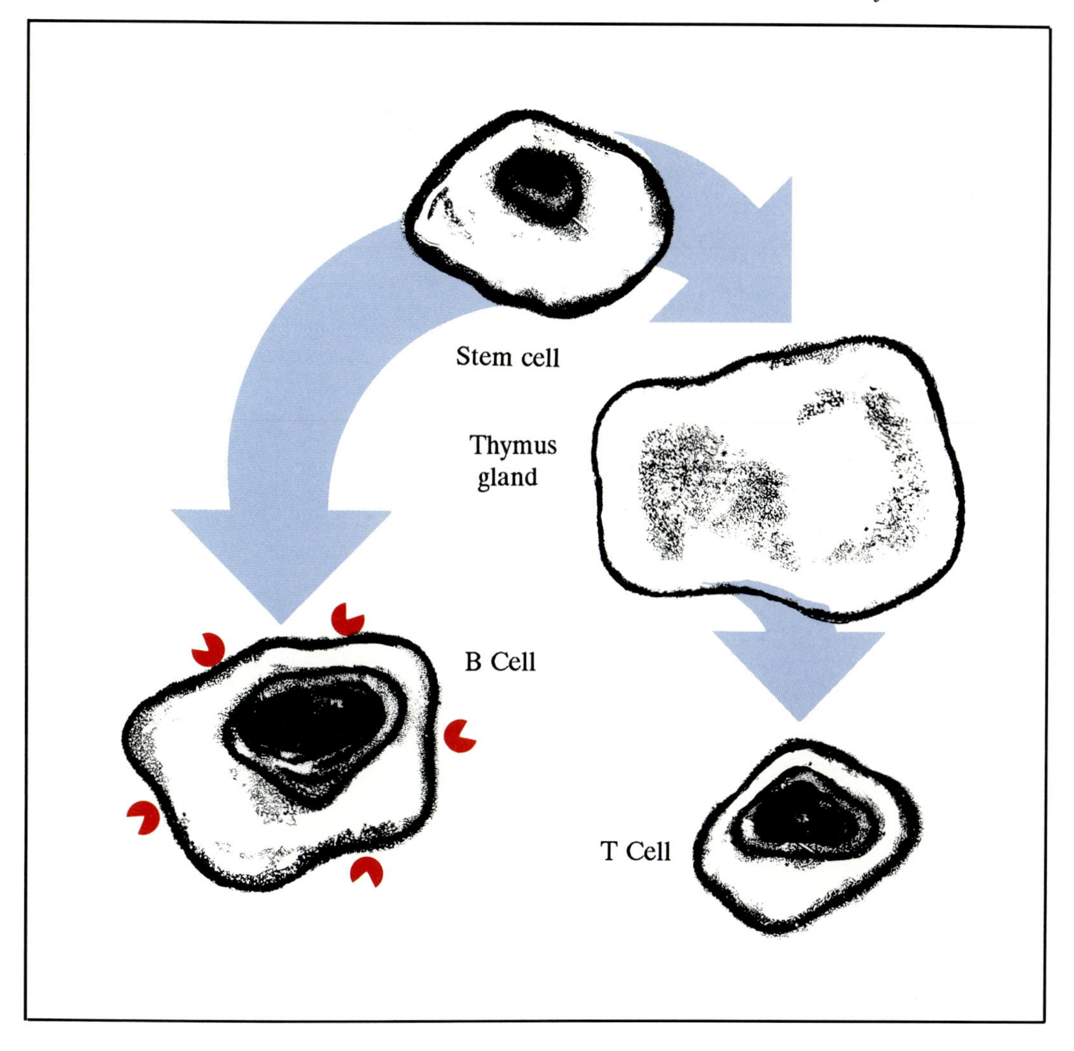

Figure 1. White blood cells called lymphocytes play a major role in the immune response. The two classes of lymphocytes, B-cells and T-cells, originate in bone marrow stem cells. While B-cells develop directly from stem cells. T-cells migrate to the thymus gland where they reach their mature state.

Unlike their sister cells, B cells (bone-marrow-derived lymphocytes do not act directly on foreign material. Instead, when stimulated by specific antigens, they produce and secrete antibodies (Figure 2). The antibodies combine with their antigens like a key fitting into a lock. All antibodies resemble each other in overall shape, yet each has unique regions that will make it fit to one antigen and not to another. This antibody production is called humoral immunity.

History of Tumor Immunology

In the early 1900s, experiments were conducted in transplanting tumors from one animal to another. The host animal's immune system would recognize the grafted tissue as "nonself" and would develop an immune response to destroy the tumor. These initial results produced great enthusiasm and a rash of attempts to develop a vaccination against tumors. But normal cells from the donor animal were also recognized as foreign and destroyed. It was not clear whether the animal was developing an immunity against foreign cells in general or against the tumor cells in particular. A lack of consistent results led to confusion and skepticism.

These problems, nevertheless, provided an impetus to the immunogenetic studies conducted during the 1950s. The existence of tumor-associated antigens was proved by using inbred mice, which share identical cell surface antigens. A normal animal could recognize these antigens as foreign and could develop an immune response to protect against later tumor challenges.

Antigens on tumor cells trigger an immune response in both B-cells and T-cells. The B-cell response may well involve interaction with helper T-cells, particularly in the early phases involving production of specific antibodies. T-cells may kill cancer cells either directly or in cooperation with other immune system components such as macrophages (Fidler, 1985) (Figure 3).

Tumor Antigens

Antigens are chemical structures that can be recog-

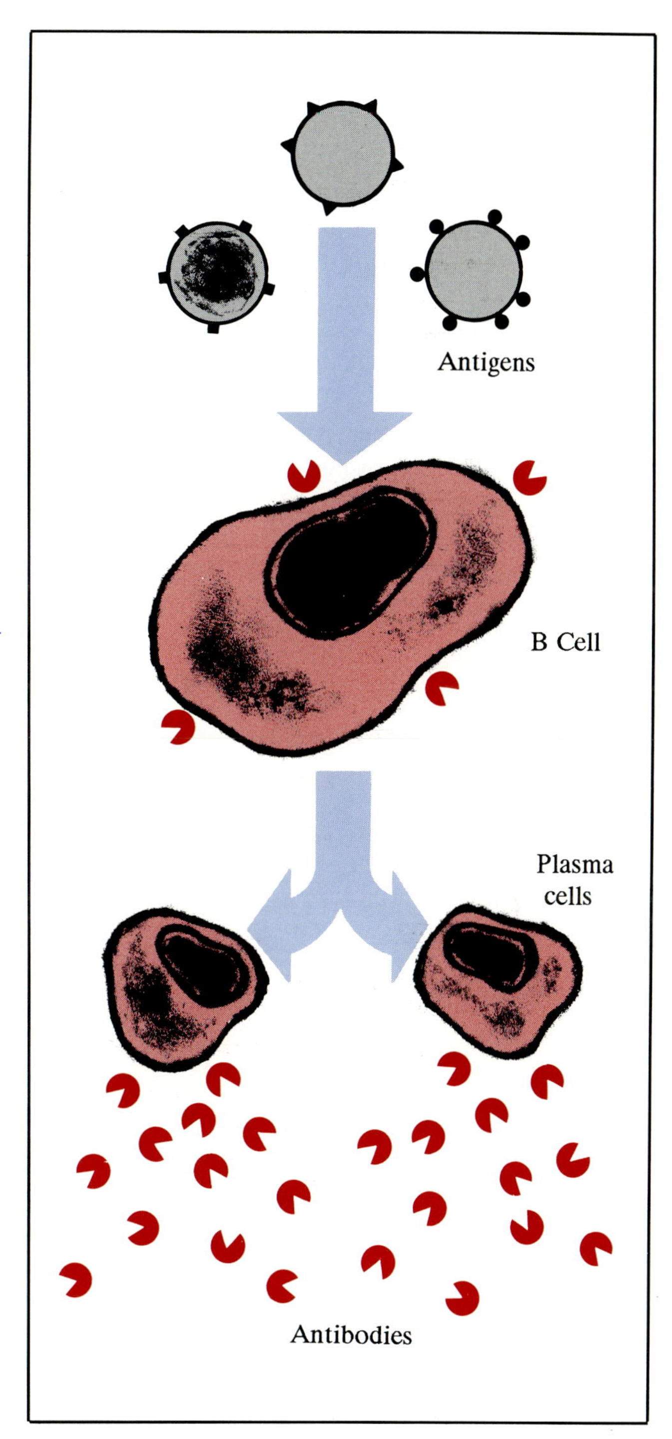

Figure 2. Antibodylike molecules on the surface of B-cells recognize and react to specific antigens. Activated B-cells then convert to plasma cells that secrete antibodies. These specific antibodies can attack and destroy foreign cells.

nized by either cellular or humoral immune response. One aim of tumor immunology has been to determine whether there are unique antigens associated with human cancers that would allow patients to recognize their own tumors as foreign. Boon and Kellerman (1977) demonstrated at the Ludwig Institute in Brussels, Belgium, that apparently nonimmunogenic tumors can be made to trigger an immune response. They treated a murine teratocarcinoma cell line with a mutagen and obtained clones that failed to grow in normal animal hosts, but did grow in immunosuppressed mice. These variant clones not only elicited a response against themselves, but also protected the hosts against the lethal parent tumor. Furthermore, the immunogenic variant produced its own antigen(s), in addition to the common antigen shared with the parent tumor cells. This result demonstrated that although tumor cells may themselves generate no immune response, they do express antigens that host defenses can recognize. Tumor cells possess many normal surface antigens, including histocompatibility antigens, tissue-specific antigens (liver, brain, pancreas, etc.), and antigens characteristic of the cancerous state.

Various changes in a cell can give rise to antigens, which in turn stimulate an immune

Figure 3. Specific molecules on the tumor cell surface act as antigens to elicit an immune response. These antigens trigger a response in both B-cells and T-cells. The T-cells can either act directly to attack tumor cells or can act as "helper" cells by interacting with B-cells to produce antibodies. T-cells can also release proteins that activate macrophages.

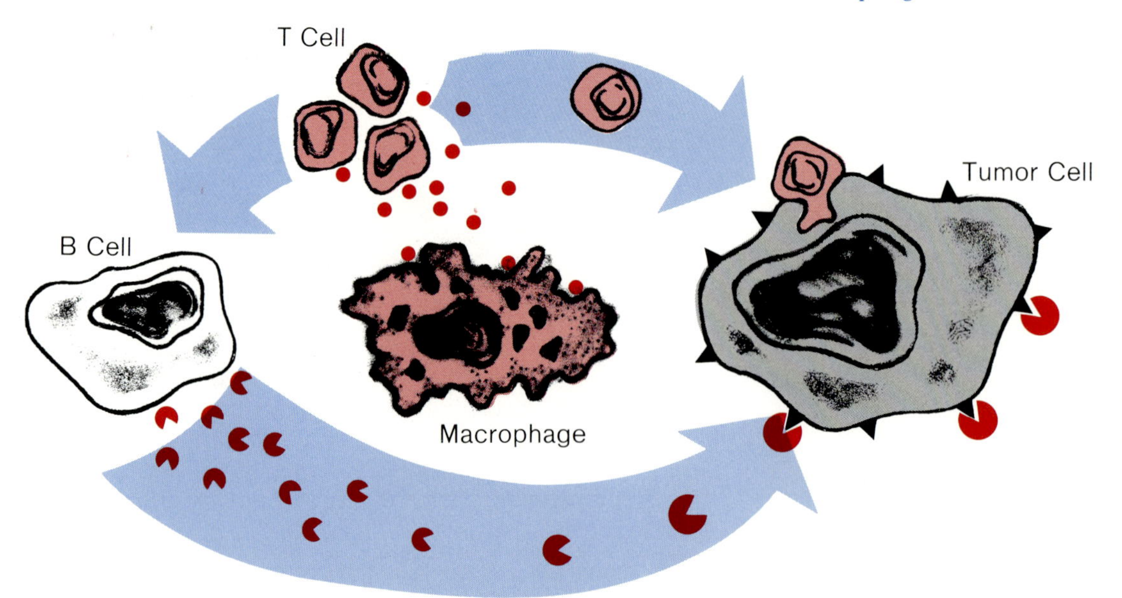

response. Although there is no single cause for the appearance of new antigens, tumor antigens provide important clues to changes involved in cancerous transformations. A number of different antigens are associated with human tumors and with various nonmalignant tissues. Among these are oncofetal antigens, surface glycolipid and glycoprotein antigens, and oncogene products. Most of these antigens are recognized more readily by the immune response of other persons or other species than by the immune response of the cancer patient. Though they do not stimulate an immune response to tumor growth, these antigens could still provide useful markers in the detection, monitoring, and treatment of human cancers.

With the advent of hybridoma technology (Kohler and Milstein, 1975), B-lymphocytes from immunized hosts could be fused with drug-selected non-antibody-secreting murine myeloma cells. These fused cell products could be cloned and propagated indefinitely. The supernatant fluids of cultures of these cloned-cell populations could yield homogeneous populations of antibodies with the desired reactivity. The many monoclonal antibodies generated in this way have led to the identification and characterization of tumor-associated antigens.

Many of these monoclonal antibodies have been used in patient diagnosis, prognosis, and treatment, as well as in clarifying the biology of cancer cells. For example, antibodies tagged with radionuclides such as iodine-131 can be injected into tumor-bearing hosts in order to locate small tumors that would be undetectable by other clinical means. Monoclonal antibodies can also be directed against tumor-associated antigens in therapy. Iodine-131-labeled antibody against antigens like carcinoembryonic antigen (CEA) produces objective clinical remissions in patients with hepatic carcinomas. Other potential therapies include antibody–drug and antibody–toxin complexes (for review, see Schlom, 1986). The antigens recognized by many of these monoclonal antibodies can be classified into three categories: oncofetal antigens, glycolipid and glycoprotein antigens, and oncogene products.

Oncofetal Antigens. Oncofetal antigens are expressed during embryonic development. Their production is generally repressed during adult life, but may be detected in regenerating or tumor-forming tissue. At least two oncofetal antigens have been well characterized: CEA (Gold and Freedman, 1965) and alpha-fetoprotein (AFP) (Sell and Becker, 1978). Although these antigens are located at the cell surface, they may be found in the blood and other body fluids after being shed from cells.

CEA is a high molecular weight, cell-surface glycoprotein usually associated with tumors of the human colon and fetal colon tissue. Early studies suggested that elevated serum levels of CEA may indicate colonic cancer. However, subsequent studies showed that CEA may be elevated in other types of cancer (such as pancreas or lung) and in various inflammatory diseases (Muraro et al., 1985). CEA measurement is particularly valuable after surgery or other therapy, to determine the degree of tumor removal and recurrence.

AFP is a serum protein normally found in high concentrations in fetal blood, but barely detectable in adult blood. This protein has attracted increasing interest because of its association with biological events such as normal and abnormal fetal development, liver regeneration, and chemically induced liver cancer. Abnormally high concentrations of AFP in maternal serum and amniotic fluid may provide valuable diagnostic clues for multiple pregnancy, abnormal transfer from fetal fluids, and decreased turnover (esophageal atresia). Serum AFP elevations occur in approximately 20% of human adults with liver injury such as hepatitis, cirrhosis, or biliary tract obstruction. Periodic determinations as a disease progresses indicate that the extent of AFP elevation reflects the degree of liver damage. Adult production of AFP by tumors is related to the tissue involved. Elevations of serum AFP are most frequently linked to yolk sac teratocarcinomas of ovary or testes and hepatocellular carcinomas. It occurs less frequently with tumors of the gastrointestinal tract and lung (Sell and Becker, 1978).

Production of these oncofetal antigens indicates that the tumor involved possesses fetal characteristics. This implies either that oncogenesis is a reversionary loss of control of gene expression or that tumors derive from stem cells retaining fetal characteristics.

Cell Surface Glycolipid and Glycoprotein Antigens. Cell surface components change during transformation of a normal cell into a malignant cell. These changes determine some of the important properties of cancer cells, including loss of contact inhibition, decrease in cell adhesion, increased growth, prolonged survival, expression of new antigens, and escape from immune destruction by the host. Cell surface carbohydrates seem to be involved in determining many of these properties. Most known tumor-associated antigens are either glycolipids or glycoproteins; in many instances, their carbohydrate structures include antigenic determinants.

In vitro and in vivo malignant transformation, induced by a variety of transforming agents, often results in changes of the cell surface sugar components of glycoproteins and glycolipids (Hakomori, 1985). Changes in glycolipid composition, resulting in either reduction or elongation of sugar chains (Figure 4), have also been observed in various human tumors. Reduction in the size of the sugar chain is probably due to lowered activity of a particular enzyme that adds sugar to peptides or other sugars.

However, some transformed clones tend to accumulate more complex glycolipid (Hakomori, 1985). The glycolipid changes seen in transformed cells have been related to growth of tumors. More recently, the quantity of G_{D3} ganglioside in human melanoma has been connected with the ability of the cells to metastasize (Schirrmacher et al., 1982). The blood of patients with certain cancers shows increased levels of tumor-associated gangliosides. For several investigators, this result implies that a

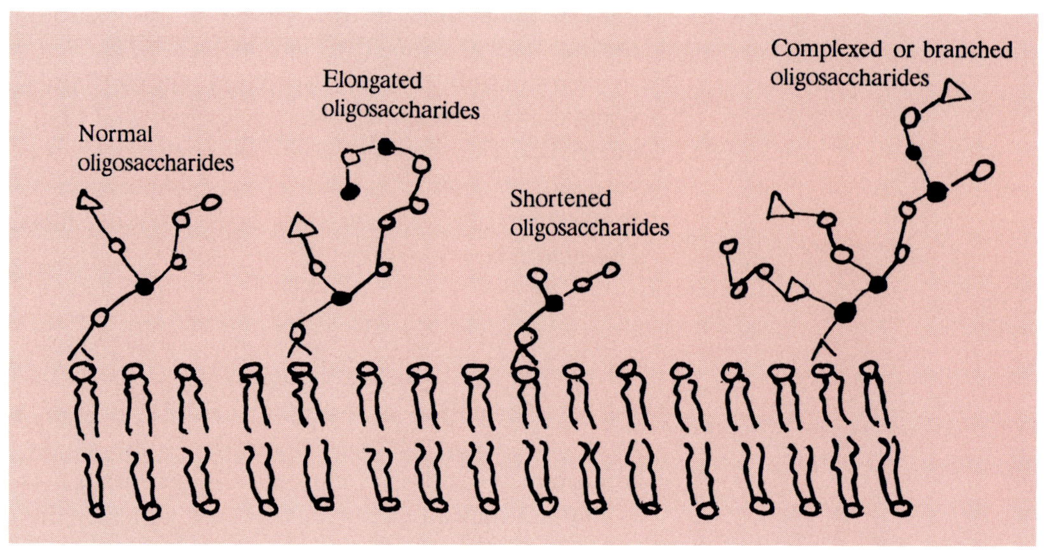

Figure 4. Changes in sugars that can affect the biological functions and immune recognition of cell surface glycolipids and glycoproteins.

high level of gangliosides may contribute to the decreased immune system function often found in cancer patients.

Blood group antigens, believed to confer individuality to cells, are carbohydrate structural determinants expressed in glycoproteins of body fluids, secretions, and membrane components. Blood group ABO antigens are widely distributed in blood cells, blood serum, endothelial cells, and epithelial cells. Many investigators have found little or no ABO activity in adenocarcinoma of the intestine and ovary, and in transitional cell carcinoma of the bladder (Limas et al., 1979). The loss of A or B determinants in these epithelial tumors results from deficiency of terminal glycosyl transferases involved in completing the respective polysaccharide chain. Measurement of these tumor-associated glycolipid and glycoprotein antigens in body fluids and tumor sections, a useful aid in tumor diagnosis, is currently the main focus of many laboratories. However, enthusiasm for these antigens as tumor markers should never undermine the need to determine the biological functions of cell surface carbohydrate antigens in tumor metastases.

Oncogene Products. DNA transfection, combined with recombinant DNA technology, has led to current concepts about the basis of human cancers. Isolated genomic cancer cell DNA was introduced into an established cell line of non-tumor-forming mouse 3T3 fibroblasts. Structurally transformed foci of cells soon arose. Foci, piled-up cells that rise above the flat configuration of normal cells in culture, result when transformed cells lose contact growth inhibition (Figure 5). Cells within the foci are transformed as DNA is incorporated from the cancer cells. This demonstrates that cancer cells contain DNA sequences (oncogenes) that are capable of transforming normal cells into cancer cells. Furthermore, these oncogenes are normal cellular genes that evolved into transforming genes as a result of mutation (for complete review, see Bishop, 1987).

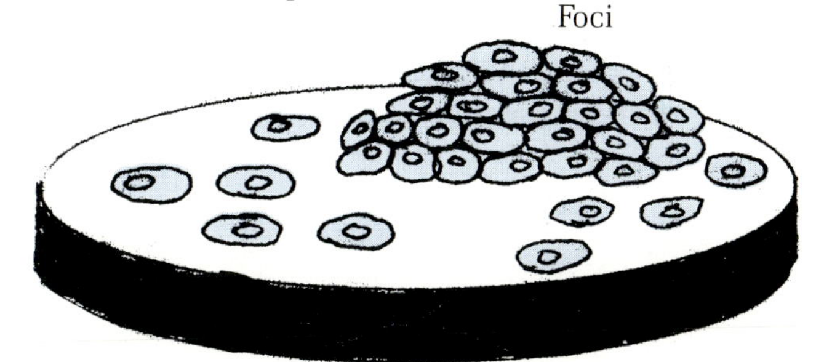

Figure 5. Transfection with DNA from tumor cells that can lead to the formation of foci.

Transition of normal cellular proto-oncogenes to cancer genes can occur in several ways, including point mutations, amplification at the gene level, over-expression at the mRNA or protein level, and chromosomal translocations or rearrangements. Several oncogene products are closely related to growth factors or their receptors: the *src* oncogene is homologous to the insulin receptor, c-*erb*-B is homologous to the epidermal growth factor receptor, and *sis* is homologous to platelet-derived growth factor. It follows that continual production of growth factors or their receptors, without the regulatory processes imposed during the normal cell cycle, could result in unregulated growth and proliferation. Specific antibodies, however, can detect abnormal quantities of these products. Therefore, oncogene-encoded molecules act as tumor antigens. Their production by tumors originating in different tissues may prove useful diagnostically or in monitoring tumor formation (Stock et al., 1987).

The frequent association of altered proto-oncogenes with cancer seems much more than coincidental. Whether oncogenes are primary factors in neoplasia or are secondary participants in the disease, future studies will clearly identify the role of these genes in the life cycle of a cell and thus allow us to understand the molecular abnormalities in cancer cells.

Immune Response to Tumors

Most known tumor antigens are too weak to elicit an intense response in the host. One explanation for the scarcity of detectable, strongly immunogenic human tumors may be that such tumors are actually rejected before they are observed. In the presence of vigilant immune surveillance, most strongly antigenic tumor cells would be destroyed. In effect, this could constitute a selection for weakly antigenic cancer cells.

Immune surveillance as a means of controlling tumor development implies a normally functioning immune system. Persons with immune deficiency diseases and patients undergoing immune suppressive therapy (for example, kidney transplant recipients) experience an abnormally high occurrence of certain tumors that are seen in patients with AIDS (acquired immune deficiency syndrome). Although they have a high incidence of Kaposi's sarcoma and B-cell lymphoma, these patients show little increased susceptibility to other human neoplasms.

Heterogeneity of Antigen Expression. Even if most cells within a tumor express potentially immunogenic antigens, a small fraction may lack them. Substantial heterogeneity in the antigens produced by a given tumor has been demonstrated with antisera and cell-mediated immunity. Heterogeneity related to different phases of the tumor cell cycle or metabolic conditions is reflected in the degree to which tumor cells build up immunity. Such cells can avoid rejection by antigenically mimicking the host's own tissues or camouflaging their surface antigens.

Antigenic Modulation. Some tumor-associated antigens will disappear from the cell surface within minutes during incubation with antibodies. When antibodies are removed, the tumor-associated antigens are again expressed on the cell surface This phenomenon is known as antigenic modulation. Although not all determinants can modulate, those that do provide a means by which the tumor cells can avoid destruction by the immune system.

Antigens Shed or Secreted. Molecules shed or released from their tumor cells can accumulate in the serum or be excreted into the urine. Serum or urine is tested for specific tumor-associated antigens in diagnostic and prognostic procedures (Liu et al., 1987). Serum levels of oncofetal antigens and tumor-associated surface glycolipid and glycoprotein antigens can increase as tumors grow. Tumor-associated antigens may be shed in substantial quantities through several mechanisms, including cell death and normal shedding of antigens from the cell surface (Black, 1980). Antibodies that recognize specific tumor-associated antigens may combine with circulating antigens and thus be prevented from binding to the tumor cells. Circulating immune complexes have been found in patients with some solid tumors, as well as in lymphoreticular malignancies (Theofilopoulos, 1982).

Immune Response Suppressors. Cancer cells themselves may act to suppress normal immune responses. Cancers of the B-cells, such as chronic lymphocytic leukemia and multiple myeloma, impair antibody formation and result in reduced function of the remaining normal B-cells. A more general form of immune impairment, involving T-cells, occurs with most types of tumors. Cancer patients show weak cellular immune responses, as indicated by skin tests with antigens such as mumps, tuberculin, and streptococcus, whereas normal individuals who have been exposed to these organisms show a strong reaction. These skin tests induce inflammatory responses that require participation of both T-cells and macrophages.

Malnutrition, a common side effect in cancer patients, is also associated with defects in cell-mediated immunity.

In addition, various inhibitory substances are either produced by the tumors or stimulated by their presence. Tumor cells can produce inhibitors of normal lymphocyte and macrophage response to chemical substances, as well as inhibitors of lymphocyte reproduction and activation. Tumors can also subvert the host's immune system in other ways. For example, the release of thromboplastin-like activity that prompts fibrin to collect around tumor cell clusters could prevent activated immune cells from attacking the tumor. Generation of low molecular weight fibrin degradation products by tumor-released proteases is another mechanism by which tumor cells could escape the immune response system, since some fibrin degradation products inhibit antigenic stimulation of lymphocytes.

Conclusion

The biological significance and function of tumor-associated antigens are becoming clear. The detection of these cell-surface molecules has led to improved methods for cancer diagnosis and therapy. Relationships between the molecular biology of oncogenes and their role in surface receptors, growth factors, and cellular regulation are just beginning to shed light into the cellular mechanisms of tumor formation. Complex immune responses and immune control mechanisms are beginning to be defined. The development of monoclonal antibodies (in which single clones of antibody-producing B-cells are immortalized and continue to produce larger quantities of specific antibodies) has extended research and clinical applications to human cancer. With the availability of monoclonal antibodies directed against tumor cells and other biological factors that can affect the cellular immune response, we can now explore new ways to augment the immune system for improved detection, treatment, and prevention of cancer.

References

Bishop, J. M. "The Molecular Genetics of Cancer." *Science (Washington, DC)* **1987,** *235,* 305.

Black, P. H. "Shedding from the Cell Surface of Normal and Cancer Cells." *Adv. Cancer Res.* **1980,** *32,* 75.

Boon, T.; Kellerman, O. "Rejection by Syngeneic Mice of Variants Obtained by Mutagenesis of a Malignant Teratocarcinoma Line." *Proc. Natl. Acad. Sci. USA* **1977,** *74,* 272.

Fidler, I. J. "Macrophages and Metastasis–A Biological Approach to Cancer Therapy: Presidential Address." *Cancer Res.* **1985,** *45,* 4714.

Gold, P.; Freedman, S. O. "Demonstration of Tumor-Specific Antigens in Human Colonic Carcinomas by Immunological Tolerance and Adsorption Techniques." *J. Exp. Med.* **1965,** *121,* 439.

Hakomori, S.-I. "Aberrant Glycosylation in Cancer Cell Membranes as Focused on Glycolipids: Overview and Perspectives." *Cancer Res.* **1985,** *45,* 2405.

Kohler, G.; Milstein, C. "Continuous Cultures of Fused Cells Secreting Antibody of Predefined Specificity." *Nature (London)* **1975,** *256,* 495.

Limas, C.; Lange, P.; Fraley, E. E.; Vassella, R. L. "A, B, H Antigens in Transitional Cell Tumors of the Urinary Bladder. Correlation with the Clinical Course." *Cancer* **1979,** *44,* 2099.

Liu, B. C. S.; Neuwirth, H.; Zhu, L. W.; Stock, L. M.; deKernion, J.B.; Fahey, J. L. "Detection of Onco-Fetal Bladder Antigen in Urine of Patients with Transitional Cell Carcinoma." *J. Urol.* **1987,** *137,* 1258.

Muraro, R.; Wunderlich, D.; Thor, A.; Lundy, J.: Noguchi, P.; Cunningham, R.; Schlom, J. "Definition by Monoclonal Antibodies of a Repertoire of Epitopes on Carcinoembryonic Antigen Differentially Expressed in Human Colon Carcinomas Versus Normal Adult Tissues." *Cancer Res.* **1985,** *45,* 5769.

Schirrmacher, V.; Altevogt, P.; Fogel, M.; Dennis, J.; Waller, C. A.; Barz, D.; Schwartz, R.; Cheingsong-Popov, R.; Springer, G.; Robinson, P. J.; Nege, T.; Brossmer, W.; Vlodavsky, I.; Paweletz, N.; Zimmerman, P. H.; Uhlenbruck, G. "Importance of Cell Surface Carbohydrates in Cancer Cell Adhesion, Invasion and Metastasis." *Invasion Metastasis* **1982,** *2,* 313.

Schlom, J. "Basic Principles and Applications of Monoclonal Antibodies in the Management of Carcinomas: The Richard and Hinda Rosenthal Foundation Award Lecture." *Cancer Res.* **1986,** *46,* 3225.

Sell, S.; Becker, F. F. "Alpha-Fetoprotein." *J. Natl. Cancer Inst.* **1978,** *60,* 19.

Stock, L. M.; Brosman, S. A.; Fahey, J. L.; Liu, B. C. S. "Ras Related Oncogene Protein as a Tumor Marker in

Transitional Cell Carcinoma of the Bladder." *J. Urol.* **1987,** *137,* 789.
Theofilopoulos, A. N. "Immune Complexes in Cancer." *N. Engl. J. Med.* **1982,** *307,* 1208.

CHAPTER 10 Biology of Cancer Metastasis

Isaiah J. Fidler and Margaret L. Kripke

Cancer has traditionally been classified as a single disease. However, it is clear now that cancer is a label for a collection of distinct, but related, diseases. Clinically speaking, all tumors can be divided into two major groups: benign and malignant. Benign tumors grow slowly, never spread, are enclosed within a fibrous capsule, and contain cells that resemble their normal precursors. Prompt diagnosis and treatment of benign tumors generally result in a cure. Malignant tumors tend to grow rapidly, invade other tissues, are rarely encapsulated, and contain many abnormal cells of different sizes and shapes (Figure 1).

Unfortunately, classifying tumors by microscopic appearance is often inaccurate and gives only a rough idea of their biological behavior. Hence, the characteristics that reliably identify tumors as malignant are invasion and metastasis.

The spread of malignant cells from a primary tumor to produce a new growth in distant organs is the most devastating aspect of cancer. Metastasis is the transfer of disease from one organ to another not directly connected to it. The process involves the release of cells from the primary tumor, dissemination to distant sites, arrest in the small blood

1420–4/88/0159$07.00/0

vessels of organs, infiltration into the stroma (connective tissue forming a framework) of those organs, and survival and growth into new tumors (Figure 2). The outcome of metastasis depends on both host characteristics and tumor cell properties; the balance of these interactions varies among tumor systems.

Our understanding of the process of tumor formation has increased considerably. However, improvement in the treatment of metastatic disease produced by the major solid tumors has been less satisfactory. Despite important advances in general patient care, surgical techniques, and supportive therapies, most deaths from cancer are caused by therapy-resistant metastases. In most patients, metastasis may have occurred by the time a primary malignant neoplasm has been diagnosed. The tumor deposits can be located in different organs or in different sites within the same organ.

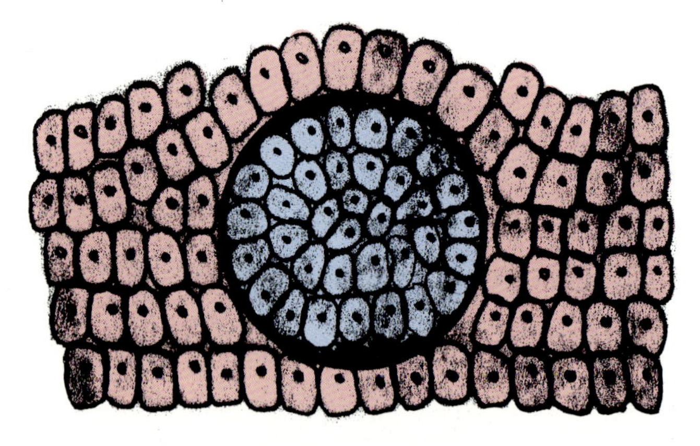

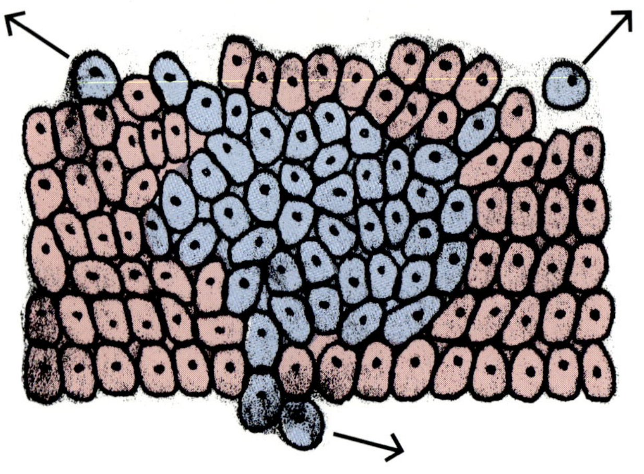

Figure 1. Benign tumors (upper) grow slowly, never spread, are encapsulated, and contain cells resembling normal precursors. Malignant tumors (lower) grow rapidly, are invasive, are rarely encapsulated, and contain many abnormal appearing cells.

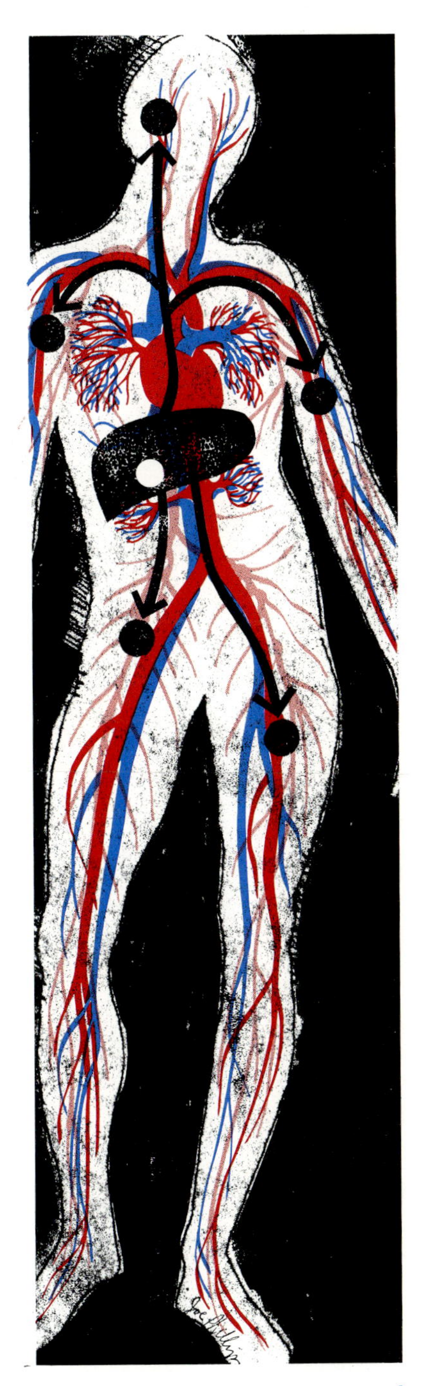

Figure 2a. Tumors can spread through direct extension, and by the lymphatic and blood systems.

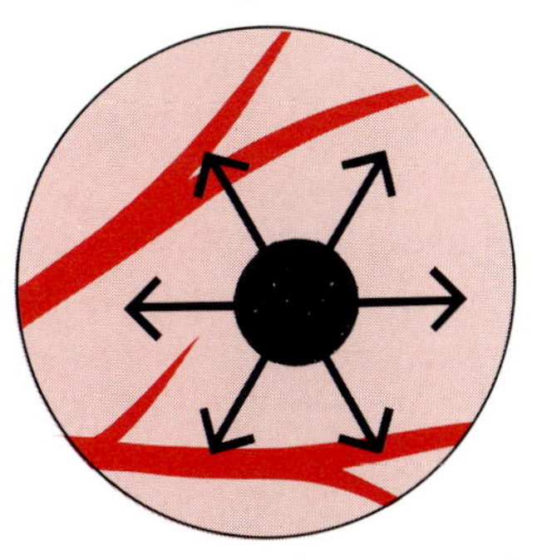

Tumor cells invade surrounding tissue

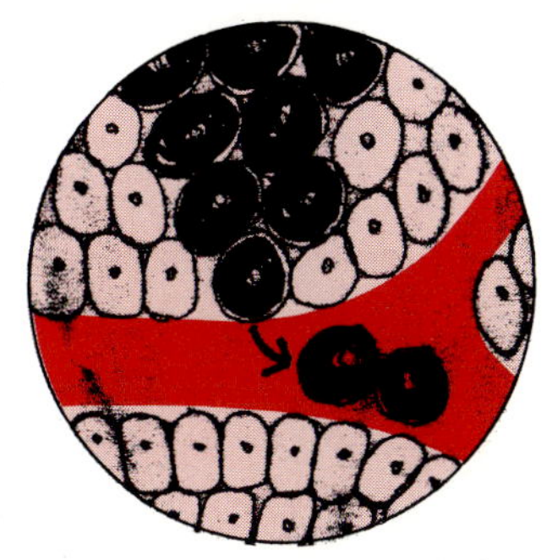

Tumor cell clumps, called emboli, are released into the circulation

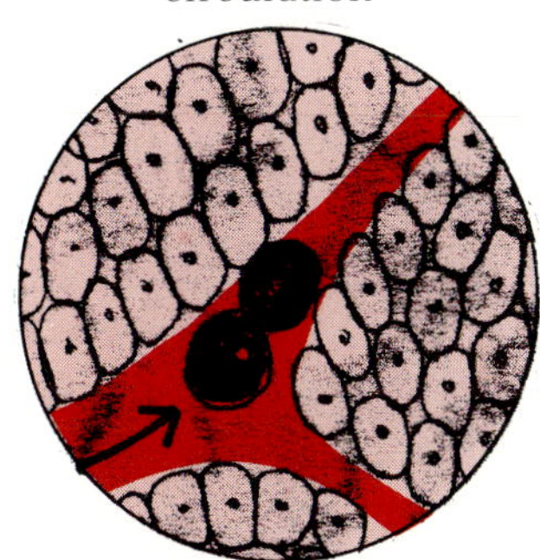

Circulating emboli are trapped in small blood vessels

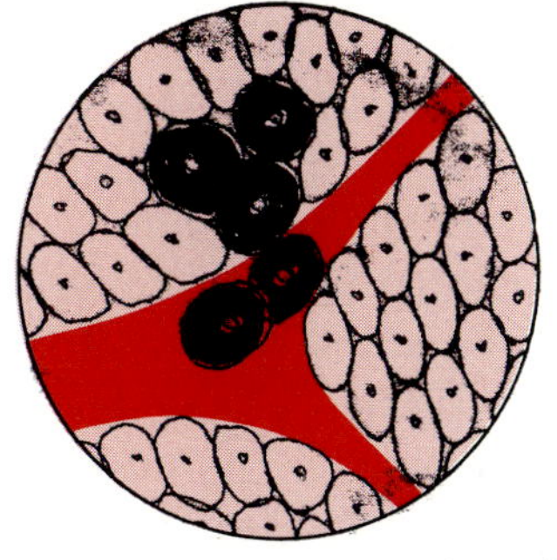

Tumors penetrate vessel walls into adjacent tissue and begin to multiply. The process may begin again

Figure 2b. The steps of metastasis.

Several aspects of metastasis exert a significant influence on the response of malignant cells to therapy. The location of a tumor will determine whether anticancer drugs will reach it in amounts sufficient to destroy the tumor cells without creating undesirable side effects. However, the biggest obstacle to the treatment of metastases is the fact that cancer cells populating both primary and secondary neoplasms are biologically heterogeneous. By the time of diagnosis, and certainly in clinically advanced lesions, malignant tumors contain cell populations that exhibit a wide range of biological characteristics such as cell surface structures, growth rate, sensitivity to various cytotoxic drugs, and the ability to further invade and spread. In addition, metastatic lesions can be large by the time they are diagnosed. A tumor mass at the lower limit of detection is approximately one cubic centemeter and may contain about one billion cells. Even destruction of 99.9% of these cells, a remarkable therapeutic achievement, still leaves one million cells to proliferate. This large residue can generate fatal, treatment-resistant, tumor cells. All these reasons make it safe to state that, short of the prevention of cancer, today's major cancer research goal is to better understand the mechanisms responsible for cancer metastasis and the generation of biological diversity, in order to design effective treatments.

Development of Metastasis

The dynamic process of metastasis can be described as a sequence of interrelated steps (Figure 2b). If the disseminating tumor cell fails to complete one of these steps, it is eliminated. Thus, malignant cells that eventually develop into metastases have survived a series of potentially lethal interactions. The outcome of this process depends on both the intrinsic properties of the tumor cells and host responses or factors. The development of a tumor can be divided into the following sequential steps:

- Metastasis begins with the local invasion of the surrounding host stroma by either single cells or clumps of cells from the primary tumor.

- Once the invading tumor cells have penetrated the blood or lymph channels, they may grow at the site of penetration or they may detach as single cells or aggregates to be transported within the circulatory system.
- Circulating clumps of tumor cells, called emboli, must survive encounters with various host immune and nonimmune defenses such as blood turbulence, lymphocytes, monocytes, and natural killer cells.
- Tumor cell emboli that survive the circulation must settle in the capillary bed of a distant organ.
- Tumor cells penetrate through the walls of these blood vessels into the organ tissue, where they multiply.
- To grow in the organ parenchyma, the micrometastases must develop a vasuclar network, and developing tumor cells must again evade the host immune system.
- When the malignancies begin to proliferate, the new tumors may shed cells into the circulation. In a short time, a small primary tumor can produce a multitude of metastases.

Mechanisms of Tumor Cell Invasion

The first step in metastasis is when tumor cells invade and infiltrate tissues surrounding the primary tumor. The tumor cells can penetrate blood or lymph vessels, or both, and achieve widespread dispersion. Recent studies have begun to reveal the mechanisms responsible for invasion of local host tissues. For example, some tumor cells can secrete enzymes that help dissolve the surrounding tissue matrix, thereby aiding the spread of tumor cells into adjacent tissues.

Continuous multiplication of tumor cells in a confined area increases pressure within a tissue, first disrupting the blood supply, later destroying normal tissue. We can compare this to penetration of soil by plant roots as they exert mechanical pressure and move along lines of least resistance.

Although multiplication of tumor cells usually

precedes invasion, it is probably not essential. The rate of cell division does not always correlate with the tumor's invasive potential. Some rapidly growing tumors, such as benign breast tumors (fibroadenomas), do not invade host tissues. In contrast, malignant breast tumors (carcinomas) may grow slowly, but invade surrounding tissue with ease. Moreover, some normal cells can also be invasive. For example, white blood cells easily invade tissues, yet they do not divide there. Many rapidly dividing normal cells, such as those in regenerating organs, remain confined to their natural environment. In short, rapid division need not be connected with the invasive process.

Individual cell motility may play a role in tumor cell invasion, though evidence for such a mechanism is only circumstantial at present (Figure 3). Certainly, tumor cells possess the internal structures necessary for active locomotion and can form cellular cytoplasmic processes that suggest spontaneous movement. Yet the inhibition of individual cells' motility can prevent invasion only in some tumors, not in all. Furthermore, increased motility is not unique to tumor cells. During fetal development, for example, normal cells migrate extensively. In adults, individual cellular locomotion is restricted primarily to white blood cells. Epithelial wound healing also involves cellular locomotion, but as a sheet of cells.

More information is available regarding the involvement of specific tissue-destructive (lytic) enzymes, such as lysosomal hydrolases and collagenases, in tumor invasion. As the host tissue is destroyed by these enzymes, an expanding tumor mass builds up pressure and blocks the blood and lymph vessels. This blockage facilitates infiltration by tumor-forming cells. Microscopic examination of the structure of tissues obtained from sites of tumor invasion reveals considerable variation in the degree of tissue damage. Many malignant neoplasms produce higher levels of lytic enzymes than benign tumors or corresponding normal tissues. Some tumors have elevated levels of lysosomal catheptic enzymes. For example, some breast carcinomas exhibit increased production of cathepsin B;

its presence suggests that this enzyme might play a role in the aggressive, malignant behavior of the tumor. Similarly, increased production and secretion of the serine protease, plasminogen activator, has been associated with the invasive behavior of some neoplasms.

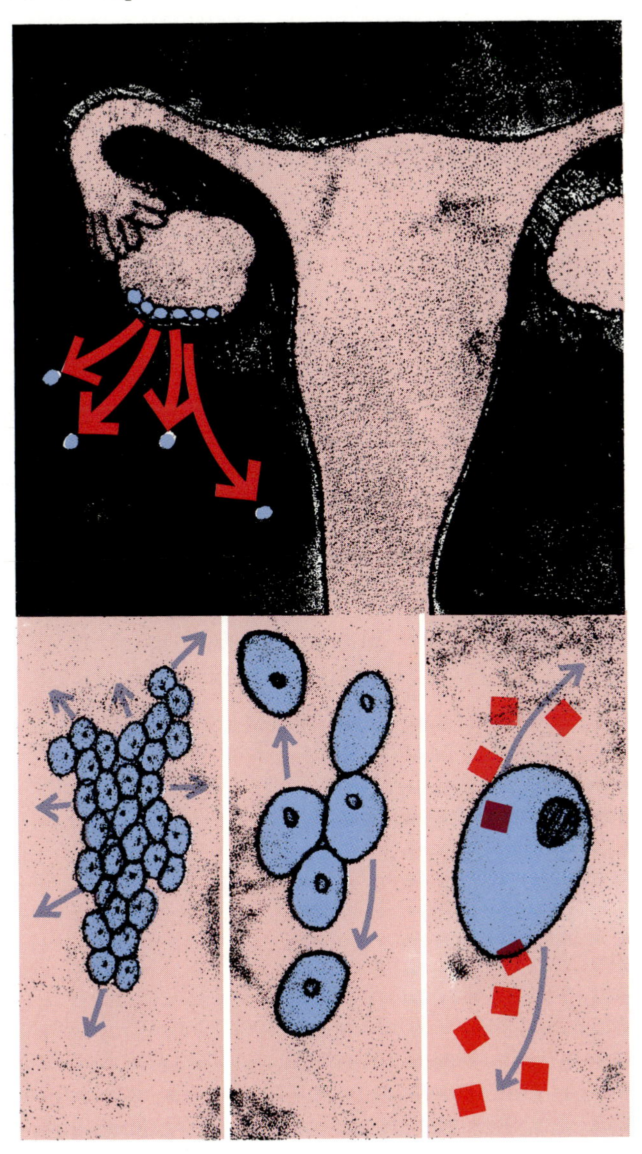

Figure 3. Malignant tumor cells of ovaries can be shed into the abdominal cavity (upper). Tumor cells may spread directly (lower left). Reduced stickiness may account, in part, for tumor cell invasiveness (lower middle). Also, secretion of proteins such as proteases might facilitate tumor cell invasion (lower right).

Penetration of blood vessels, during both invasion and extravasion, is crucial in metastasis. In order to survive, tumor cells that invade blood vessels or exit from capillaries of distant organs in which they have lodged must penetrate the basement membrane. Dissolution of this membrane in areas next to arrested tumor cells suggests enzymatic action. Basement membranes and connective tissues contain four major groups of molecules: collagens, elastin, glycoproteins, and proteoglycans. Relative quantities vary among different tissues and basement membranes. These extracellular matrix constituents are stabilized and organized by a variety of protein–protein and polysaccharide–protein interactions. In turn, the interactions can become destabilized by degradative enzymes.

The collagen and proteoglycan components are favored sites of tumor cell attachment and destruction of basement membranes. Metastatic tumor cells can show preferential attachment to the major collagen class of basement membranes, type IV collagen. These cells often secrete high amounts of an enzyme that decomposes type IV collagen specifically. Metastatic tumor cells also produce high amounts of enzymes that can cleave the major proteoglycan of basement membranes, heparin sulfate proteoglycan. Thus, these enzymes appear to be excellent markers for highly metastatic cells.

Finally, normal host responses to injury can inadvertently aid tumor cell invasion. Actively growing tumors may induce a local inflammatory reaction, causing white blood cells to accumulate in the tumor area (Figure 4). These blood cells can release hydrolytic enzymes that destroy host tissues and thereby enhance tumor cell invasion.

Mechanisms of Cancer Spread

Most malignant neoplasms eventually spread to distant organs. The number, size, and distribution of secondary tumor growths depend on many factors, including the length of time the patient has had the disease. Tumor cells may spread by one or more of three major routes: direct extension or transplantation, through the lymphatic system, and through the bloodstream (Figure 2).

Direct Extension. Tumors that grow in or invade body cavities can release freely traveling cells that

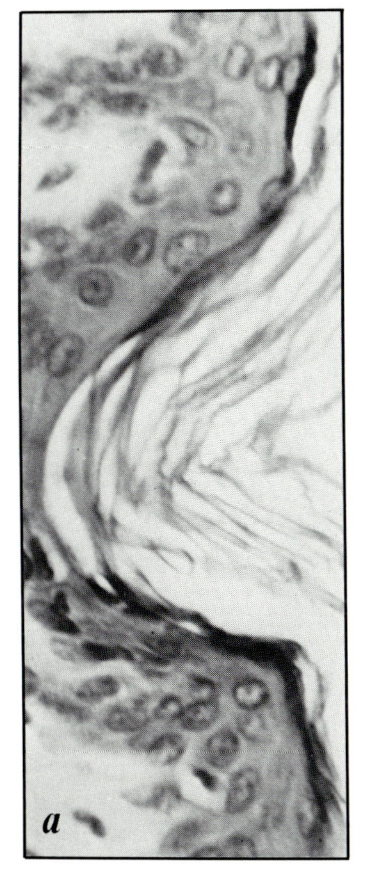

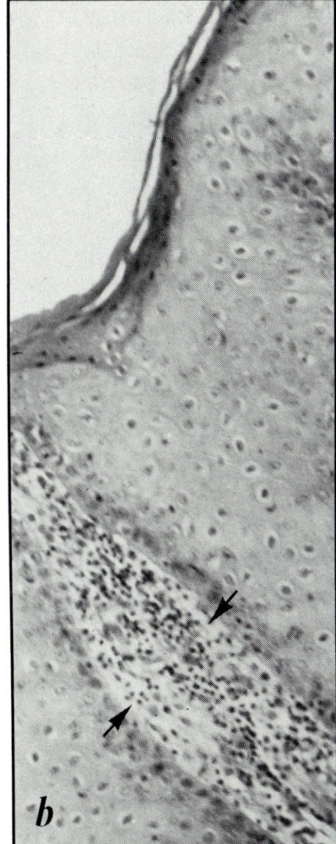

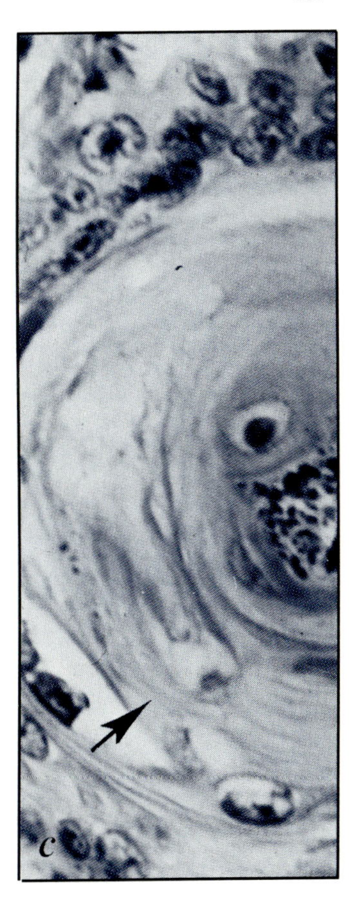

Figure 4. This series shows normal and malignant skin cells as they appear under the microscope. a. Normal skin cells including stratified squamous epithelium. Note the regular alignment of cells and the even strand of the protein, keratin. b. Squamous cell carcinoma of skin shows irregularities such as the surface layer penetrating downward. Inflammation sets in, and lymphocytes (arrows) gather. c. Another example of squamous cell carcinoma of skin shows relatively developed cells which still form keratin. However, instead of regular strands—see 4a—keratin is in bizarre locations (arrow). All 400 ×.

seed the surfaces of distant organs and develop into new tumors. For example, malignant tumors of the ovary often shed cells into the abdominal cavity (Figure 3). Tumors of the central nervous system, though highly invasive, rarely metastasize to organs outside the nervous system. They apparently spread either by direct extension or through cerebrospinal fluid.

Lymphatic System. The lymphatic system is a common path for the spread of many neoplasms (Figure 4). Tumor cells frequently infiltrate a draining lymph node and spread from there to distant lymph nodes; they sometimes bypass a chain of regional lymph nodes in the process. Lymph is a fluid that collects in lymphatic capillaries, flows into larger vessels, passes through the lymph nodes, and enters the bloodstream through ducts in the neck.

In 1860, Virchow compared the lymph node to a "packed charcoal filter". This basic concept, that lymph nodes act as purely mechanical traps, has since been expanded. For example, lymph nodes may initiate and maintain host immunity. However, their role in controlling tumor spread is still unclear.

Tumors first spread in the lymphatic system by clotted cells called emboli. Tumor emboli can be trapped in the first lymph nodes encountered, or they can bypass them to form distant metastases. Malignant cells can readily move between blood and lymph channels, a fact that suggests that these circulatory systems are inseparable with regard to the spread of tumors.

Blood Stream. Widespread dissemination of tumor cells results from penetration of blood vessels, lymphatics, or both. Malignant cells frequently penetrate thin-walled capillaries, but rarely invade artery or arteriole walls, which are rich in elastin fibers. This resistance to invasion is not necessarily dictated by mechanical strength alone. Connective tissues possess protease inhibitors that may block the enzyme-dependent invasion process. Malignant tumors do not produce their own blood vessels, but induce the growth of new capillaries from host tissue by releasing tumor angiogenesis factor. A defective endothelium and increased permeability can aid in penetration of these vascular channels. Once tumor cells have penetrated into blood vessels, they can be passively carried away. Alternatively, they may grow at the penetration site and later release tumor emboli into the circulatory system. The appearance of tumor emboli signals the development of tumor vascularization.

The generation of emboli is probably a continuous process. Most malignant tumors have a well-established blood supply with multiple thin-walled vessels that can rapidly release many tumor cell emboli. A sudden change of pressure in the veins, such as occurs during a cough, could lead to momentary blood turbulence and the release of a shower of emboli. Diagnostic procedures or routine manipulation of primary tumors can also cause a sudden increase in the number of circulating tumor cells. Such emboli can be isolated from the circulatory system long before metastases form. For example, tumor cells appear in rat blood as early as 24 hours after tumors are implanted into their muscles. However, if researchers amputated the legs containing the tumors within six days after implantation, no metastases formed. Clearly, most tumor emboli do not survive to develop into new growths. Since most cells released into the bloodstream are eliminated rapidly, the mere presence there of tumor cells does not in itself constitute metastasis.

The rapid death of most circulating tumor cells is probably due to traumatic blood turbulence, although the isolated embolus can interact with a variety of blood components. Tumor cells can aggregate with each other (homotypic aggregation) or with host cells such as platelets and lymphocytes (heterotypic aggregation). The formation of such multicellular emboli improves the survival rate of circulating tumor cells (Figure 5).

Tumor cells must become entrapped in the capillary bed of distant organs before secondary tumors can form. Although the structural aspects of tumor cell arrest have been studied extensively, relatively little is known about the dynamics of the process. The capillary basement membrane, exposed in the normal and continuous process of shedding endothelial cells, may allow adhesion of tumor emboli. After platelets adhere to damaged areas, their degranulation can cause further loss of endothelial cells and the attachment of additional tumor emboli or platelet–tumor cell emboli. Fibrin deposits are often found around an arrested tumor embolus. Increased tendency for blood to coagulate, commonly seen in cancer patients, could be related to the high levels of thromboplastin found in certain tumors. It has been shown recently that some neoplasms produce high levels of procoagulant-A activity, which can directly activate the clotting process. A reduced rate of blood flow could lead to increased trapping of circulating tumor cells and increased survival of cells already trapped. The use of anticoagulants in the treatment or control of metastasis is based on such factors.

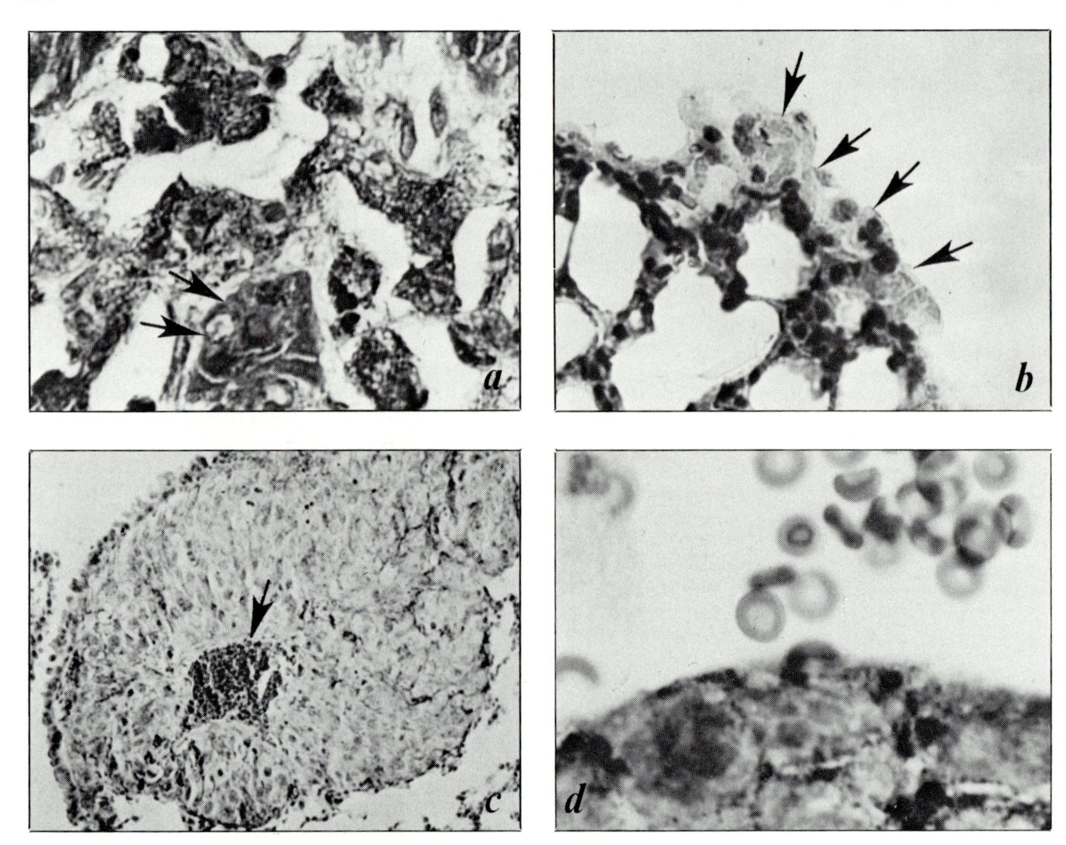

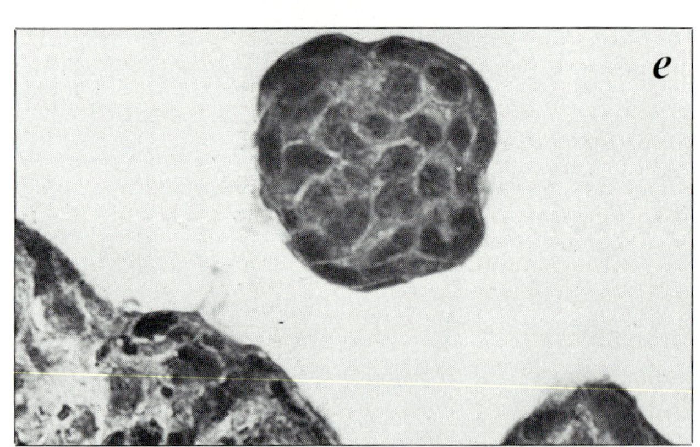

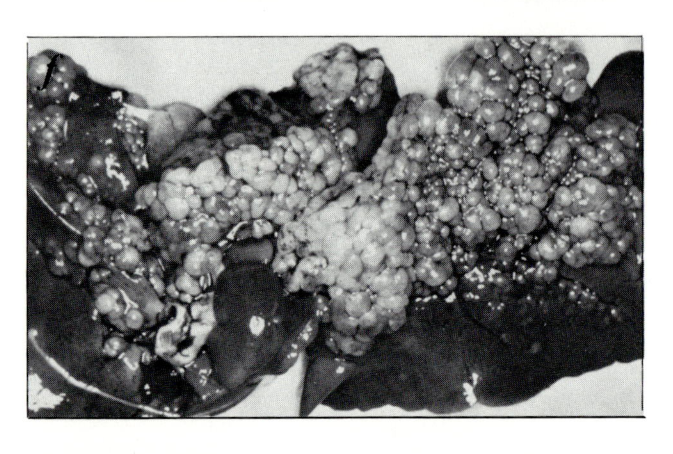

Figure 5. Examples of invasion and metastasis. a. Malignant melonoma cells shown invading oral cavity tissues. The malignant cells (arrow) have destroyed normal tissue, leaving the lower layers disarrayed, 400 ×. b. Mouse lung tissue with metastases induced by experiment. Some tumor cells cluster (arrows) to begin colonies, 400 ×. c. Mouse lung metastasis originating from malignant melanoma. One large tumor mass has a well-developed blood vessel (arrow), 400 ×. d. Malignant melonoma in the lung shown growing into a blood vessel containing disc-shaped red blood cells, 600 ×. e. Tumor cell clumps, such as these from the lung, can recirculate, arrest, and form new metastases, 400 ×. f. A guinea pig was injected with carcinoma cells that made their way to the liver (shown here) where they formed many tumor colonies, 10 ×.

Extravasation of arrested tumor cells seems to occur by mechanisms comparable to those that permit invasion. Tumor cells grow and destroy the vessel sheltering them before erupting into the surrounding tissue. To be able to grow in the organ parenchyma, metastases must both induce the development of a vascular network and evade the host immune system.

In both humans and animals, metastases are found frequently in some organs (such as the lung and liver, Figure 6), but rarely are found in other organs (such as spleen, skeletal muscle, cartilage, or thyroid). The major reason may be that most of the venous blood drains into the lungs and liver. Metastases are also seen in the bones of many human cancer patients.

Do tumor cells tend to reach one particular organ, or do they go everywhere, but develop only in selected organs? Evidence suggests that both mechanical and local "soil" factors (to use the plant root analogy) determine whether metastases will develop after tumor emboli lodge in an organ. Factors such as clot formation, localized injury, and oxygen supply can help determine attachment of tumor cells. For example, liver metastases could be aided by direct injury, altered blood flow, or other diseases that damage the liver, such as chronic inflammation (cirrhosis).

Tumor cells begin multiplying soon after new blood vessels begin to nourish them. Studies described by Folkman in this volume suggest that tumor cells actually stimulate the host to furnish a capillary network. A diffusable molecule, called the tumor angiogenesis factor, is involved. When tumors have attained a certain size, they may give rise to others, the so-called "metastasis of metastases".

During the process of dissemination, tumor cells are susceptible to destruction by host-specific (lymphocytes, antibodies) and nonspecific (natural killer cells, macrophages) defense mechanisms. Many investigators have studied the possibility that innate host immune mechanisms can control the spread of cancer. In practice, at least three major issues need to be addressed: (1) the heterogeneous nature of malignant neoplasms, (2) the intrinsic antigenicity/immunogenicity of metastatic tumor cells, and (3) the ability of the host to recognize and destroy susceptible tumor cells. Related experimental or clinical approaches include increasing tumor cell antigenicity; interfering with immunoregulation (by eliminating suppressor cells), or specific or nonspecific immune stimulation; transfusion with effector cells (*NK* cells, *T* cells); inhibiting tumor growth and/or metastasis by administration of monoclonal antibodies and immunoconjugates; or destroying metastatic cells by in situ activation of host defense cells, such as lymphokine-activated killer cells and macrophages. Although all of these approaches have been efficacious in at least one animal tumor model, their clinical application awaits confirmation.

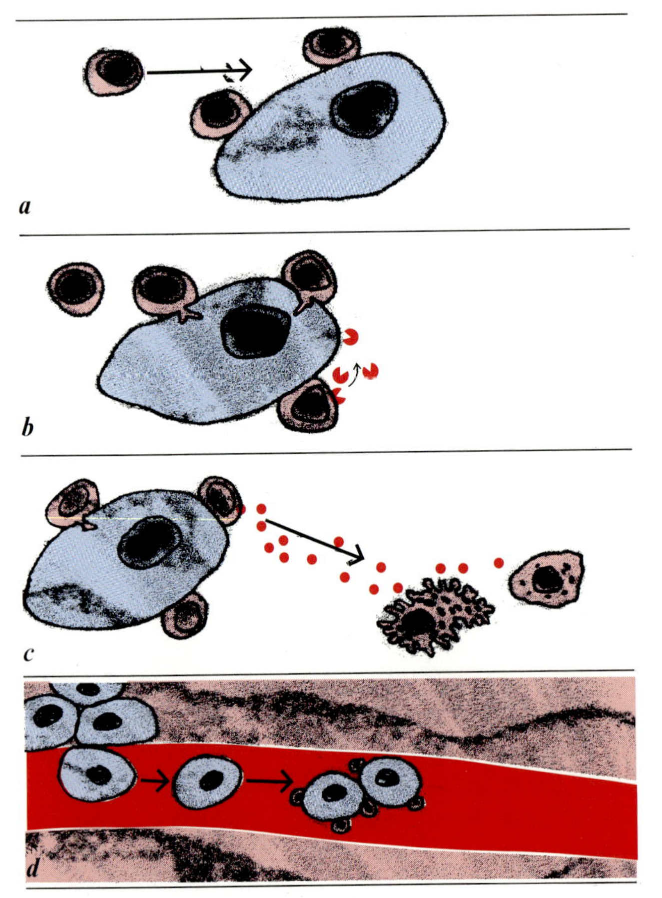

Figure 6. Clumping of tumor cells by host lymphocytes might actually aid the spreading and survival of tumor cells. a. Lymphocytes meet and attach to tumor cells. b. Lymphocytes attack tumor cells and may produce antibodies (red). c. Lymphocytes secrete proteins that attract macrophages. d. Clumping by lymphocytes of tumor cells in the blood stream may help their spread. The photograph (below) shows actual clumps of mouse malignant melonoma cells with lymphocytes. 400 ×.

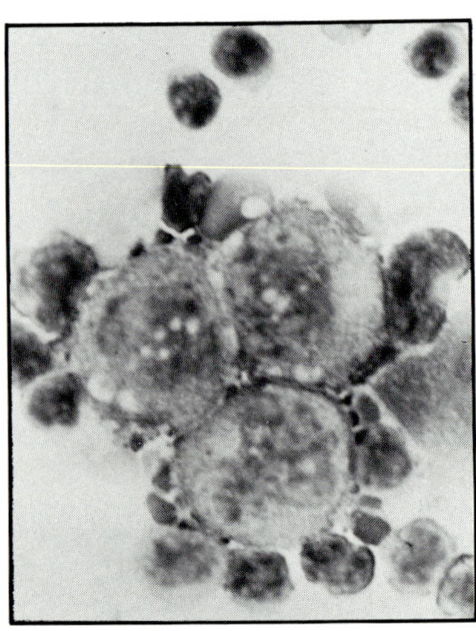

Heterogeneous Nature of Metastatic Neoplasms

It was first assumed that any malignant cell was capable of spreading within the body. In the last few years, however, this concept has changed dramatically. When we injected radiolabeled tumor cells into the circulation of mice, we observed that fewer than 1% of them were still viable 24 hours after entry into the circulation (Fidler, 1970). Fewer than 0.1% of the tumor cells placed into the circulation produced metastases. Such observations prompted us to question whether the 0.1% of the circulating cells responsible for the development of metastases represented random survival, or whether we were charting the selective survival and growth of cells endowed with special properties. In other words, can all cells growing in a primary neoplasm produce metastases? Do only specific cells possess properties that enable them to survive the potentially destructive pathway from the primary tumor to metastases?

The concept of heterogeneous neoplasms that contain subpopulations of cells with different biological properties is not new. Almost a century ago the English pathologist, Paget, analyzed autopsies of a large number of patients with breast cancer. He concluded that the nonrandom pattern of metastasis was not due to chance. Rather, some tumor cells ("seeds") traveling by vascular routes had an affinity for growth in the environment provided by certain organs ("soil"). Metastases developed only when the "seed and soil" were matched. Similarly, recent experiments have shown that many experimental tumors can produce site-specific metastases in rodents. A modern version of the "seed and soil" hypothesis would have to include three important principles: (1) the process of metastasis is not entirely random; (2) neoplasms are not uniform entities, but contain cells with varying metastatic capabilities; (3) the outcome of metastasis depends on properties of both the tumor cells and the host. The balance of these contributions varies among tumors arising in different tissues, and even among tumors of similar structural types arising in different patients.

Cells with varying metastatic properties have been isolated from individual parent or primary tumors, a finding indicating that the cells in a primary tumor are not all equal in their ability to disseminate. Two general approaches have been used to isolate populations of cells that differ in their metastatic capacity. In the first, tumor cells are selected in vivo (Figure 7). The cells are implanted into mice, metastasis is allowed to occur, and the metastatic lesions are harvested. Cells that are recovered from the metastases can be expanded in cell culture or can be used immediately to repeat the process; the cycle is repeated several times. The behavior of the ultimately recovered cells is compared with that of the cells of the parent tumor to determine whether the selection process produced cells with enhanced metastatic capacity.

In the second approach, cells are selected for the enhanced expression of a characteristic believed to be of importance in one or another step of the metastatic sequence. Then they are tested in the appropriate host to determine whether metastatic potential has been increased or decreased. This method has been used to examine whether properties as diverse as resistance to lymphocytes and antibodies, adhesive ability, lectin resistance, and invasive capacity (Figure 8) are important in metastasis.

One obvious criticism of these studies is that the surviving isolated tumor cell may have arisen as a result of adaptive rather than selective processes. We provided the first experimental demonstration of the metastatic heterogeneity of neoplasms in 1977 (Fidler and Kripke, 1977), in work with a mouse melanoma. Using a modification of the fluctuation assay devised by Luria and Delbruck, 1943 for analyzing the origin of microbial mutants, we showed that individual cells from the parent tumor varied dramatically in their ability to metastasize. Single cells were isolated from the parent tumor and grown into individual cell lines, or clones. The cloned cell lines, each originating from a different tumor cell, were tested for their

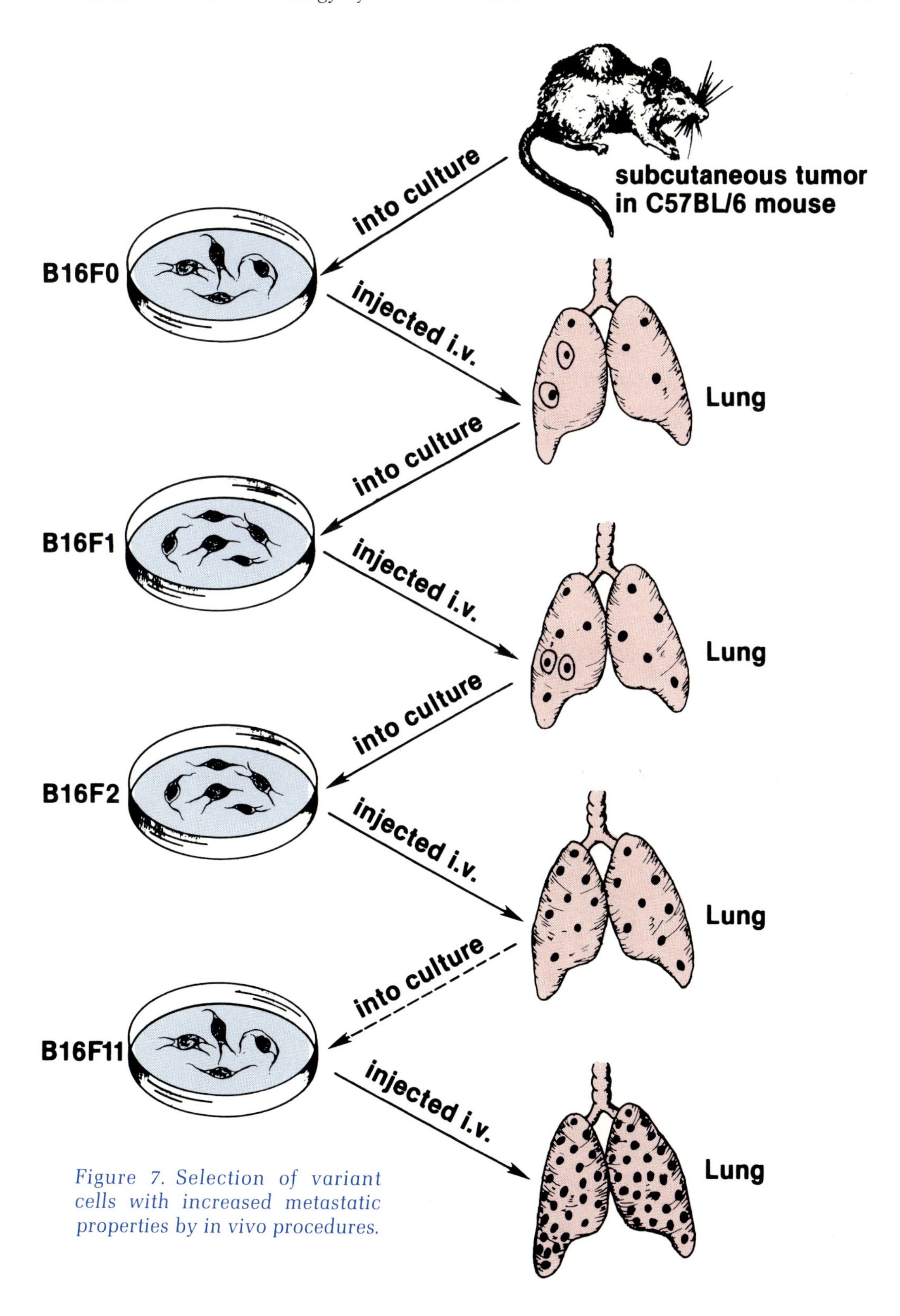

Figure 7. Selection of variant cells with increased metastatic properties by in vivo procedures.

ability to form pulmonary nodules when injected into syngeneic mice. Some clones produced many metastases; others produced very few (Figure 9). Control procedures demonstrated that the diversity did not result from the cloning procedure. These studies demonstrated that subpopulations of tumor cells with differing metastatic potential exist within the parent tumor. This finding has since been confirmed in many laboratories, using a wide range of experimental animal tumors of different histories and origins. Moreover, by using young athymic mice that are unable to reject transplants of foreign tissues, we demonstrated recently that several human tumors also contain subpopulations of cells with widely differing metastatic properties (Fidler, 1986).

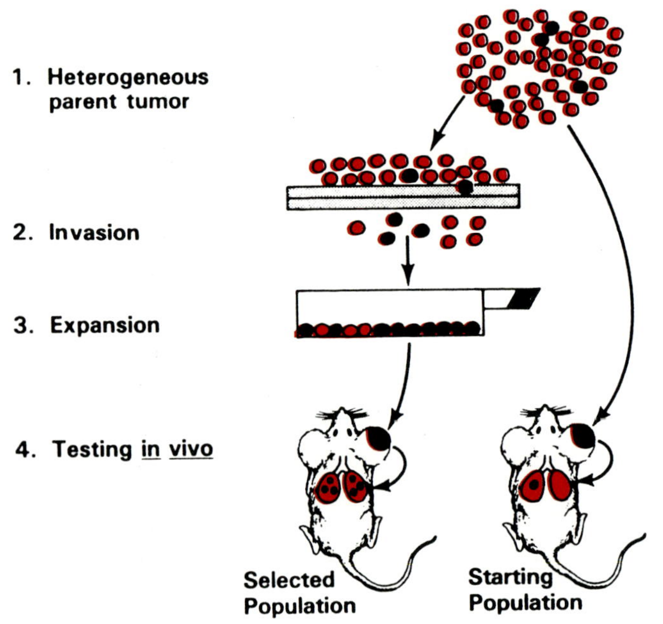

Figure 8. In vitro selection technique for highly invasive tumor cell variants that also exhibit increased metastatic capability.

Mechanisms for Generation of Biological Diversity in Neoplasms

The mechanisms by which cancer cells diversify are not fully understood. Some tumors may develop from several simultaneously transformed cells, and therefore contain different cell populations. How-

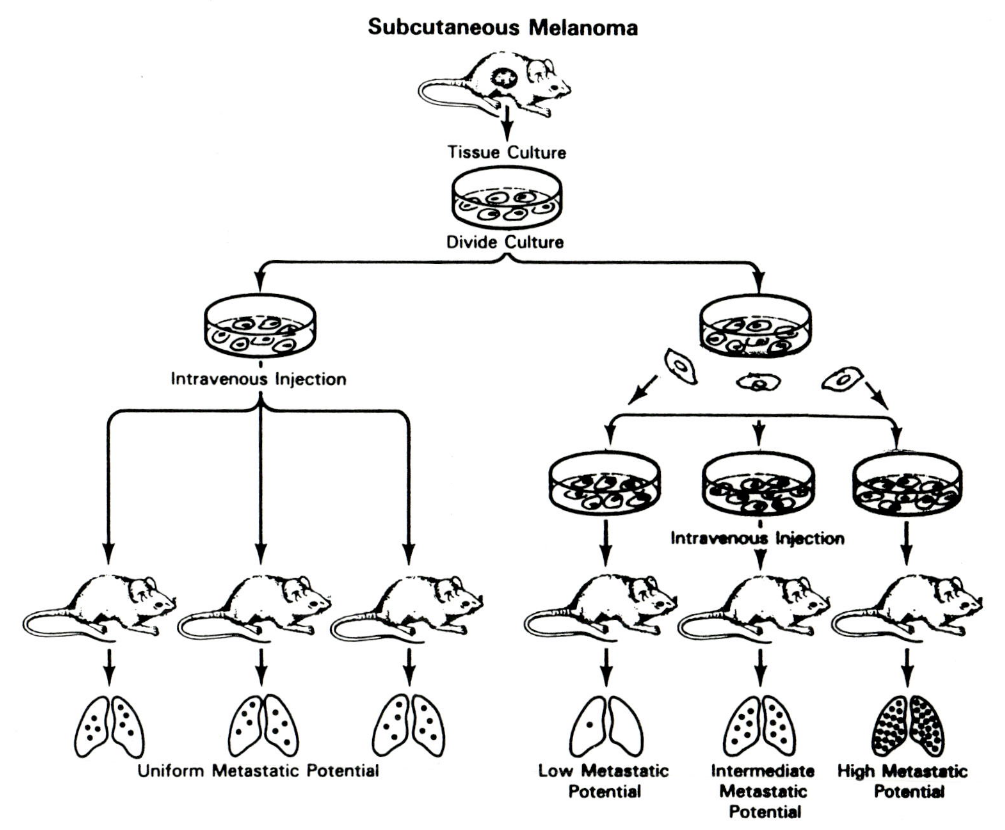

Figure 9. Design of the experiment that demonstrated that tumors are heterogeneous for metastasis, and that metastatic subpopulations of cells preexist within the parent neoplasm.

ever, most human tumors result from the proliferation of a single cell. The cellular diversity within tumors originating from a single cell probably develops by tumor progression and evolution. That is, variant tumor cells arise that differ from the parent cell. If a cell survives it will multiply, eventually generating its own new variants. Ultimately, this process creates a tumor containing numerous subpopulations of cancer cells that are related, but not identical.

The idea that malignant cells appear as the result of progressive changes was first introduced by Leslie Foulds (1954). The concept was refined by P. C. Nowell (1976), who suggested that tumor progression could result from a genetic instability of tumor cells. This hypothesis also predicts that emerging variants will display increasing genetic instability, with even more rapid emergence of the next generation of variant cells. To test this hypothesis, we examined the stability of metastatic clones and the rates of mutation of paired metastatic and primary cloned lines isolated from four different mouse neoplasms (Cifone and Fidler, 1981). In all of these experiments, the rate of spontaneous mutation was significantly higher in metastatic cells than in nonmetastatic cells.

Evolution and progression occur both in primary neoplasms and in metastatic lesions. Multiple metastases in the same patient, even within the same organ, frequently are heterogeneous with regard to many biological characteristics. This cellular diversity may be a consequence of the process of tumor evolution, or it may result from the nature of tumor cell spread. Recent studies (Fidler and Talmadge, 1986) have shown that many metastases develop from a single cell (or progenitor). Moreover, different metastases in the same organ can originate from different progenitor cells. This fact may account for some of the biological diversity among various metastases, although other sources of diversity are also probable.

Variant tumor cells are generated faster in homogeneous tumors than in tumors containing large numbers of subpopulations (Fidler and Talmadge, 1986). This fact may explain why new variants are generated so rapidly in metastases, which generally arise from one or several tumor cells. Diversity is at a minimum early in the growth of metastases. Because this condition is conducive to a high rate of variant generation, heterogeneous metastases rapidly develop. Why variant cells arise within tumors and what factors control their rate of appearance are not known at present.

Implications for Cancer Therapy

However it is generated, the heterogeneity of metastatic tumors has important implications for their study and treatment. For example, the structure, activity, and drug sensitivity of cells obtained from a primary tumor do not necessarily reflect the properties of the cells populating its metastases, or even cells in different regions of the same metastasis.

Successful therapy of cancer metastases, therefore, requires the development of approaches that can circumvent the heterogeneity problem. Given the extraordinary amount of cellular diversity within many tumors, a single anticancer drug, or any other treatment used alone, will probably not be capable of killing all of the cancer cells in a malignant tumor and all of its metastases. Consider a treatment that destroys most, but not all, of the tumor cells in a cancer. It may inadvertently be stimulating the surviving cells into regenerating new variants, perhaps more malignant and resistant to treatment.

During the last few years, we have concentrated our efforts on systemically activating macrophages to kill tumor cells. For reasons that are not completely clear, macrophages discriminate accurately between neoplastic and normal cells. They preferentially lyse tumorigenic cells, regardless of their individual characteristics. Moreover, although we can select tumor cells that are resistant to other components of the immune system, so far we have been unable to select tumor cells that are resistant to killing by macrophages. We have been devising methods to enhance the killing activity of macrophages. Encouraging successes in laboratory settings are stimulating our hope that this type of biological therapy may eventually prove useful in the clinic (Fidler, 1985).

Summary

Metastasis is a complex, highly selective process that depends on the interplay of host factors and tumor cell properties. Characteristics of tumor cells, such as their surface structures, adhesive capacities, motility, and enzyme secretion, appear to be important. To establish metastases, tumor cells must complete all steps involved in the process. Enhanced performance by a cell in one step of the process does not compensate for an inability to complete a subsequent step. Interruption of the sequence at any stage can prevent the production of a clinical, visible metastasis. Metastasis is governed by rules and patterns. The ultimate goal for biologists interested in this process is to understand the common features of its pathogenesis. Because primary malignant tumors are not composed of cells with uniform metastatic behavior, we must carefully evaluate the adequacy of many tumor systems now used for study and devise new approaches for the treatment of metastatic disease.

Acknowledgment

This chapter is based on research supported in part by funds from the R. E. "Bob" and Vivian L. Smith Foundation.

Suggested Reading

Cifone, M. A. "Increasing Metastic Potential is Associated with Increasing Genetic Instability of Clones Isolated from Murine Neoplasms." *Proc. Natl. Acad. Sci U.S.A.* **1981,** *78,* 6949–6952.

Fidler, I. J. "Metastasis: Quantitative Analysis of Distribution and Fate of Tumor Emboli Labeled with ^{125}I-5-iodo-2′-deoxyuridine." *J. Natl. Cancer Inst.* **1970,** *45,* 773–782.

Fidler, I. J. "Macrophages and Metastasis-A Biological Approach to Cancer Therapy: Presidental Address." *Cancer Res.* **1985,** *45,* 4714–4726.

Fidler, I. J. "Rationale and Methods for the Use of Nude Mice to Study the Biology and Therapy of Human Cancer Metastasis." *Cancer Metastasis Rev.* **1986,** *5,* 29–49.

Fidler I. J.; Hart I. R. "Biological Diversity in Metastatic Neoplasms: Origins and Implications." *Science (Washington, DC)* **1982,** *217,* 998–1003.

Fidler, I. J.; Kripke, M. L. "Metastasis Results from Preexisting Variant Cells Within a Malignant Tumor." *Science (Washington, DC)* **1977,** *197,* 893–895.

Fidler, I. J.; Poste G. "The Cellular Heterogeneity of Malignant Neoplasms: Implications for Adjuvant Chemotherapy." *Semin. Oncol.* **1985** *12,* 207–221.

Fidler, I. J.; Talmadge, J. E. "Evidence That Intravenously Derived Murine Pulmonary Melanoma Metastases Can Originate from the Expansion of a Single Tumor Cell." *Can. Res.* **1986,** *46,* 5167–5171.

Foulds, L. *Neoplastic Development* ; Academic: London, 1975; Vol. 2.

Heppner, G. "Tumor Heterogeneity." *Cancer (Philadelphia)* **1984,** *214,* 2259.

Liotta, L. A. "Tumor Invasion and Metastasis—Role of the Extracellular Matrix: Rhoads Memorial Award Lecture." *Cancer Res.* **1986,** *46,* 1–7.

Luria, S. E.; Delbruck, M. "Mutations of Bacteria from Virus Sensitivity to Virus Resistance." *Genetics* **1943,** *28,* 491–511.

Nicolson, G. L.; Poste, G. "Tumor Cells Diversity and Host Responses in Cancer Metastasis." *Curr. Probl. Cancer* **1982,** *7,* 4–83.

Nowell, P. C. "The Clonal Evolution of Tumor Cell Populations." *Science (Washington, DC)* **1976,** *194,* 23–28.

Poste, G.; Fidler, I. J. "The Pathogenesis of Cancer Metastasis." *Nature (London)* **1979,** *283,* 139–146.

Tarin, D.; Price, J. E.; Kettlewell, M. G.; Souter, R. G.; Vass, A.; Crossley, B. "Mechanisms of Human Tumor Metastasis Studied in Patients with Peritoneovenous Shunts." *Cancer Res.* **1986,** *44,* 3584–3592.

Weiss, L. *Principles of Metastasis* ; Academic: New York, 1985.

CHAPTER 11 Angiogenesis

Judah Folkman

Capillary blood vessels thinner than a hair deliver blood to every organ in the body. These tiny blood vessels rarely grow, except during the process of angiogenesis, in which new capillary blood vessels proliferate. Angiogenesis can occur, for example, in a healing wound for a few days until the wound begins to close, and then the blood vessels return to their resting state. Tumors, however, can induce the continuous growth of new capillary blood vessels to converge upon the tumors themselves. In fact, progressive tumor growth seems to be dependent upon the daily production of hundreds of such new blood vessels. In this chapter we will examine the reasons that tumor growth is angiogenesis-dependent, as well as the mechanisms by which tumors induce angiogenesis.

Normal Cell Growth

Normal cells throughout the body are neatly packed together. Their level of crowdedness is approximately 100 million cells per cubic centimeter (10^8cells/cm^3). This cell density is very efficient for certain functions such as protection against infection, production of vital proteins, and communication between cells. It is inefficient, however, for the delivery of nutrients and oxygen to

1420–4/88/0183$06.00/0

cells and the removal of waste products. The body solves this delivery problem by interposing myriads of capillary blood vessels between cells. In some organs such as the liver, capillaries are so numerous that each liver cell is contiguous to a capillary blood vessel. In other tissues, the distance to the nearest capillary may be two or three cell widths, but it is never greater than the optimum diffusion distance for oxygen, approximately 150 μm. (A micrometer is one millionth of a centimeter, and the average width of different cell types is 20–50 μm). This relationship between cell density and available capillaries is precisely maintained in each tissue in all animal species. It can be most easily observed in the retina, where cells live in discrete layers. For example, in the bat and guinea pig, the retina is 150 μm thick and has no capillaries of its own that is, it is *avascular*). Although it is avascular, it is thin enough so that its cells can thrive on the oxygen and nutrients from the capillaries belonging to the tissue just beneath it. In the seal and rabbit, the retina is 350 μm thick and is vascularized by multiple layers of capillaries. It is not understood how this precise ratio between tissue cells and the capillary blood vessels is maintained.

The total number of cells in any given tissue or organ is also held fairly constant, despite the fact that, in some tissues, old cells are dying or shedding and new cells are being born to replace them, all at a rapid rate. The human bone marrow normally produces 10 billion new cells per hour (10^{10}) to replace both red and white blood cells removed from the circulation because of age or other losses. The entire population of bone marrow cells is replaced in approximately five days. The mucosal cells that line the intestine, the most rapidly proliferating cells in the body, have a turnover time of approximately three days. Tissues that face the external environment and contain cells that line the skin, gut, airway, and genito-urinary tracts generally maintain the highest rates of cell turnover. Interestingly, more than 90% of cancers in human adults arise from these rapidly replicating tissues. Cancers of the colon, breast, and lung are some examples.

Cells that do not directly "see" the outside environment (e.g., those in liver and kidney) have much slower turnover times, measured in months or years.

Cells Lining the Blood Vessels–Endothelial Cells

In contrast to the rapidly dividing cells of the intestine and bone marrow, *endothelial* cells, which line the inside of the blood vessels, are among the most slowly dividing cells in the body. Their turnover time is measured in years (in some tissues 10–30 years). During angiogenesis, however, endothelial cells can be stimulated to proliferate almost as rapidly as bone marrow cells, with a turnover time of a few days.

Normal or Physiologic Angiogenesis

Angiogenesis occurs normally or physiologically in females, but rarely, if ever, in males. There is a brief onset of capillary blood vessel growth in the ovary at the time of ovulation and in the uterus at the end of menstruation. Angiogenesis is also essential to the formation of the placenta. Males may go through an entire lifetime without evidence of angiogenesis unless they are wounded or suffer a severe heart attack (myocardial infarction). Thus, when angiogenesis does occur in a male, it is usually the result of a disease process.

Abnormal or Pathologic Angiogenesis

Abnormal or pathologic angiogenesis can occur in both males and females. In deep wounds or after surgical operations, capillaries begin to grow rapidly. This growth ceases after the wound closes. Unabated capillary growth appears to be the basis of a variety of diseases found in almost every branch of medicine. For example, capillary blood vessels growing abnormally in the eye cause blindness in premature babies, as well as in diabetic adults. In arthritis, newly growing capillaries can invade the joint cartilage and destroy it. In many skin diseases such as psoriasis, abnormal capillary growth is the dominant feature. The cause of abnormal capillary growth in these pathological states is not completely understood; however, it is becoming clear that many diseases, previously thought to be unrelated, can now be considered as angiogenesis-dependent diseases. In these situations, abnormal capillary growth is the principal pathologic feature, and it is possible that a treatment that could control abnormal angiogenesis could also ameliorate or control the disease itself.

The Beginning of a Tumor

Of all the diseases dominated by abnormal capillary growth, *neoplasia* is associated with the most intense and persistent angiogenesis. Angiogenesis is not necessary for the initiation of a tumor nor even for the earliest stage when a few new tumor cells are just beginning to proliferate. Angiogenesis does appear to be an essential component, however, of progressive growth of a solid tumor, of the release of metastatic cells to remote organs, and of the increasing array of serious symptoms that tumors produce in their hosts.

A tumor cell arises from a normal cell, but does not necessarily grow faster than its normal counterpart. It is a common misconception that the major difference between normal and neoplastic cells is that the neoplastic cells grow more rapidly. In fact, the opposite is often true. Normal mucosal cells of the colon, for example, divide much faster than malignant cells in a cancer of the colon. A fundamental difference between tumor and normal cells is that tumor cells continue to grow under crowded conditions that would inhibit the growth of normal cells. Normal liver cells would not grow beyond a cell density of approximately 10^8 cells/cm^3, but a liver tumor may reach a density of 10^9 malignant cells/cm^3.

Tumor cells continue to proliferate despite crowding because they seem to disregard physical information that alerts most cells to the extent of crowding. This information is generally received by cells as changes in their shape brought on by the pressure of neighboring cells. Cells are elastic. When they are uncrowded or sparse (e.g., when migrating into an open wound), they spread out. A fully spread cell may cover a surface area of up to approximately 3000 μm^2 and assume a pancake or spindlelike appearance. In contrast, a maximally crowded cell can be squeezed or rounded into a surface area of only $400 \mu m^2$. Most normal cells can proliferate when they are in the spread configuration (if the appropriate growth factors are present), but not in the small rounded configuration. In contrast, tumor cells can grow whether spread or rounded. In other words, tumor cell growth is largely independent of cell shape, although the degree of this independence varies with tumor type and the duration of the tumor.

It is not understood how the nucleus of a normal cell knows that the cell's shape is sufficiently rounded (or compressed) to shut off DNA synthesis. Recent experimental evidence suggests that this information is transmitted to the nucleus by the degree of tension in certain components of the cytoskeleton; cytoskeletal tension increases with spreading. It is also unclear what changes take place in the nucleus when information about cell shape reaches it. A recent report (Jiang and Schindler, 1988) demonstrates that access of specific mitogenic growth factors to the nucleus is tightly regulated in normal cells by cell shape—cell spreading increases the access and cell rounding shuts it off. In tumor cells, the specific growth factors have access to the nucleus even when the cell is fully rounded, that is, under the most crowded conditions.

Another characteristic of tumor cells that permits them to keep growing in a crowded population is their greater capacity for anaerobic metabolism. *Anaerobic metabolism* is an ability of tumor cells to live at slightly lower oxygen levels than normal cells. These two abnormalities, unresponsiveness to cell shape signals and an increased capacity for anaerobic metabolism, permit the formation of a tiny tumor mass, on the order of about a million cells and a few millimeters in diameter. However, tumor growth cannot usually continue beyond this point unless new capillary blood vessels are induced to grow into the tumor. Once tumor angiogenic activity is turned on, there is almost no further limit to the size of the tumor.

These observations can be assembled into a portrait of how a new tumor starts and progressive tumor growth occurs. Carcinoma of the cervix and melanoma of the skin provide the most detailed picture of tumor growth because both tumors are visible throughout most of their existence.

Basement membrane, a continuous thin sheet throughout the body, separates epithelial cells from underlying blood vessels, as well as other connective tissue cells such as fibroblasts. Basement membrane, which can be seen best by electron microscopy, contains mainly collagen, laminin, fibronectin, and heparan sulfate. These large molecules help to form a thin porous structure to which epithelial cells can attach. In most tissues, basement membrane is only 400 Å (0.04 μm) thick. In the kidney glomerulus this membrane is about 0.12 μm thick, and it is even thicker in the cornea. The normal epithelial cells of the cervix that are capable of dividing form a single layer on a sheet of basement membrane just like epithelial cells in other organs (e.g., skin and intestine) (Figure 1a). When an epithelial cell in the cervix divides, one daughter cell remains attached to the basement membrane, where it is capable of further cell division. The other daughter cell is squeezed up into the next layer of cells, where it can no longer divide. Cells in these upper layers will eventually die and be shed. Incidentally, the cells in the upper layers have a smaller rounded shape, which may contribute to their lack of further growth.

An early sign of malignancy in the cervix occurs when cells that rise into the upper layers of the cervix also begin to proliferate (Figure 1b). The accumulation of a small nest of tumor cells, many of which are undergoing mitosis, is called in situ carcinoma.

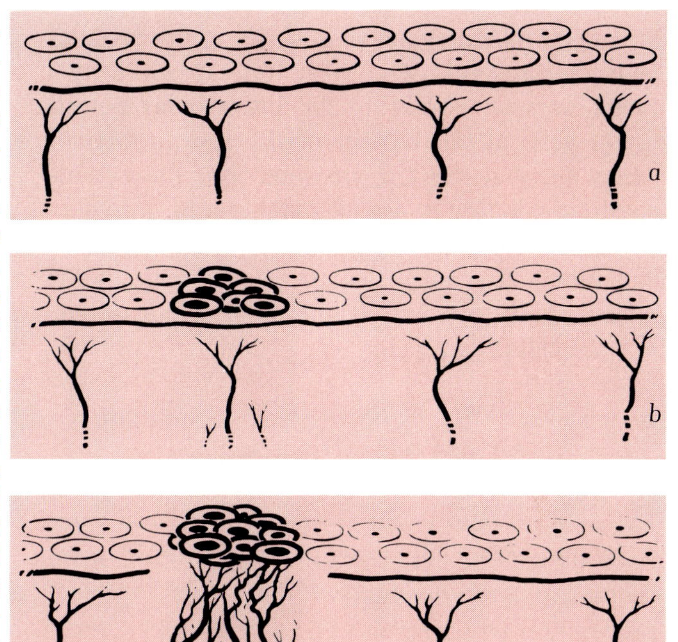

Figure 1. (a) Normal arrangement of epithelial cell layers separated by basement membrane from underlying blood vessels and connective tissue. This diagram represents skin, cervix, intestine, and other epithelial compartments. (b) The prevascular tumor. An example of early in situ carcinoma. Tumor cells lie entirely above the basement membrane and have not been penetrated by new capillaries. In this phase, tumor thickness is limited to the distance over which oxygen and nutrients can diffuse from the capillary blood vessels beneath the basement membrane. (c) The vascularized tumor. Angiogenesis coincides with a local breakdown of basement membrane. Rapid tumor growth follows. Metastasis is possible at this stage.

At this stage all of the tumor cells lie above the basement membrane and are thus living by simple diffusion of nutrients and oxygen from the capillaries beneath the basement membrane. Tumor cells cannot pile up indefinitely. The thickness of the in situ tumor is limited to a few millimeters or less. For skin melanoma, the limit of tumor thickness is known more precisely, 0.76–0.9 mm. A possible explanation for this restricted tumor thickness is that tumor cells at the top of the pile are so far removed from the nearest capillary that, even with the advantage of anaerobic metabolism, they are at the limit of oxygen diffusion. Theoretically, tumor cells could grow indefinitely over the whole basement membrane as a single cell layer, spreading away from the original nest of tumor cells. It is not understood why they rarely (if ever) do this, but it is known that tumor cells secrete specific growth and attractant factors for cells of their own kind. These factors may lead to a pattern of clustered tumor cell growth.

The Prevascular Tumor

In situ carcinomas may exist in a *prevascular* phase for several years with little or no further tumor expansion. During the prevascular period, tumor cells at the bottom of the pile proliferate, while those at the top are shed into the outside environment. (Many are still viable and, for carcinoma of the cervix, these cells are detected by the Papanicolaou smear). The total tumor population remains fairly constant at approximately a million or more cells (Figure 1b).

The Vascularized Tumor

The in situ tumor enters the *vascular* phase when it is penetrated by new capillaries that grow through the basement membrane. The onset of angiogenesis appears to coincide with local lysis of the basement membrane beneath the tumor. As capillaries breach the basement membrane, the tumor also protrudes through the membrane and surrounds the capillaries in a cylindrical fashion (Figure 1c). It is currently thought that tumors make specific enzymes that can locally destroy the basement membrane. Experimental evidence also suggests that tumor cells prefer to settle themselves near new capillary sprouts. In most cases, rapid tumor growth begins soon after vascularization.

The vascularized tumor extends vertically into the deep tissues beneath the basement membrane, where it continues to induce the proliferation of new capillary blood vessels. The enlargement of a vascularized tumor may be extremely rapid. We have observed certain animal tumors that remained static for months during the prevascular phase, but were able to enlarge 16,000 times their original volume within two weeks of becoming vascularized.

During the vascular phase, the tumor may begin to shed cells into the circulation. These wandering tumor cells can settle in distant organs and form new tumors, called *metastases*. For the human skin melanoma, a black tumor that may arise from an earlier mole, the prevascular phase remains small (0.76 mm thick or less) and almost never sends out metastases. If the tumor is surgically removed at this stage, the patient will usually live a normal life span (Figure 2). In contrast, the

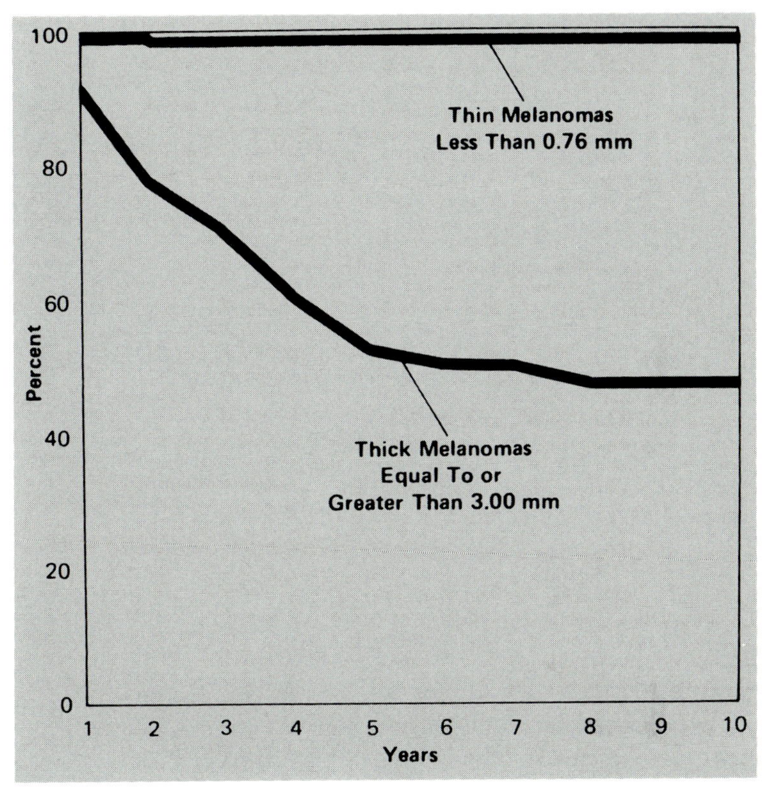

Figure 2. Percent survival (vertical scale) over 10 years (horizontal scale) of patients in whom a skin melanoma was excised. In those patients whose tumor was less than 0.76 mm thick, survival was 99%. In those patients in whom the melanoma was equal to or greater than 3.00 mm thick, the survival rate was approximately 50%. Our experimental evidence suggests that the thin (0.76-mm) melanoma is prevascular and that the thick melanoma is vascularized. (Reproduced with permission from Cady. Copyright 1987, American Cancer Society.)

vascularized tumor rapidly thickens as it begins to grow in three dimensions. Even if it is removed surgically, some tumor cells will already have escaped, and many patients with these larger tumors will die of metastases (Figure 2). There is increasing evidence that other epithelial tumors such as cancer of the colon, breast, and bladder behave similarly to carcinoma of the cervix or melanoma by evolving through a prevascular and vascular phase (Figure 3).

Tumors that arise from within organs, such as the liver, also exist for a time in the prevascular state. They are small, usually not more than a few millimeters in diameter, and are surrounded by normal cells and normal capillaries. The tumor cells at the center of these nodules may suffer the same deficit of oxygen and nutrients as those that were on top of the pile in the cancer of the cervix; they are too far removed from the nearest capillary. After new capillaries penetrate these tumors, there is rapid growth.

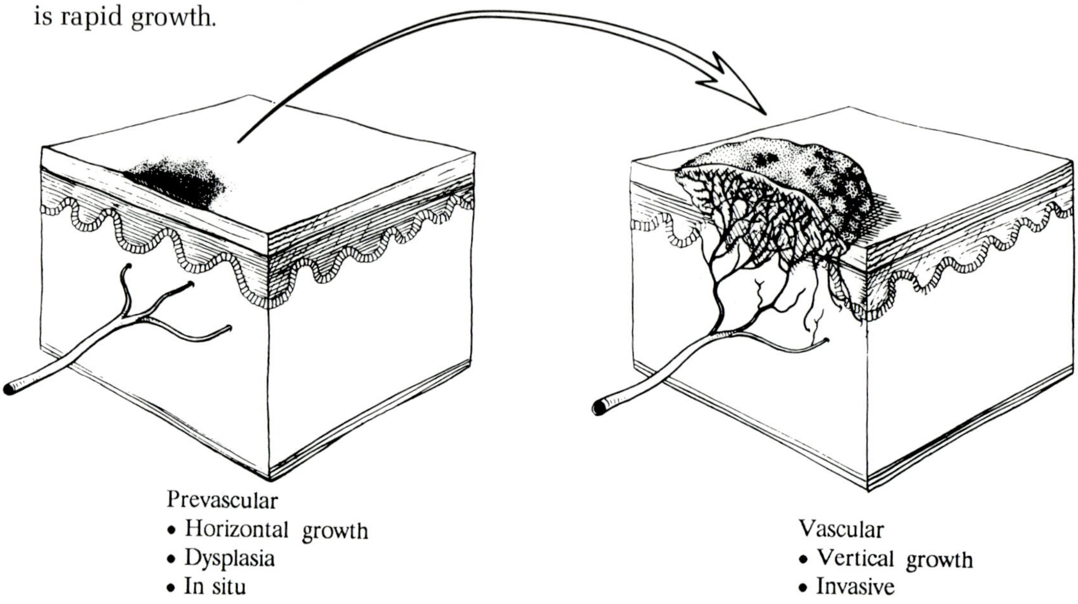

Prevasculár
• Horizontal growth
• Dysplasia
• In situ

Vascular
• Vertical growth
• Invasive

Figure 3. Diagram of an epithelial tumor in the prevascular phase (left) and after its transition to the vascularized phase (right). This model could represent a carcinoma of the cervix, bronchus, breast, colon, or bladder, or any other tumor arising in an epithelial compartment, including a skin melanoma.

Initiation of Angiogenesis

The mechanisms by which tumors induce angiogenesis have been studied for at least a decade, but are still not well understood. Most tumors can release chemical signals that stimulate the formation of new capillary blood vessels (Figure 4). At least four different angiogenic proteins that are secreted by tumors have been discovered so far. For each protein the amino acid structure has been elucidated and the gene has been identified. Although each protein appears to function in a different way, the end result is the production of new capillaries that grow toward the tumor. One of the most potent and ubiquitous of these angiogenic proteins is fibroblast growth factor (FGF). When it was first discovered more than 10 years ago, it was known only to stimulate the growth of fibroblasts. It was subsequently found to be a potent angiogenic factor, and in 1984 it was completely purified.

Many tumors appear to use additional methods to induce angiogenesis, analogous to a back-up system for eliciting the inward growth of new capillary blood vessels. Some tumors, for example, release chemicals that attract macrophages. These cells are normally preoccupied with the destruction of invading bacteria, but they also participate in many other processes, including wound healing. Macrophages move into fresh wounds and secrete angiogenic factors. One of these is fibroblast growth factor; another is tumor necrosis factor; and a third has not yet been fully characterized. When large numbers of macrophages are attracted into a tumor, they contribute to the total angiogenic activity being produced by the tumor.

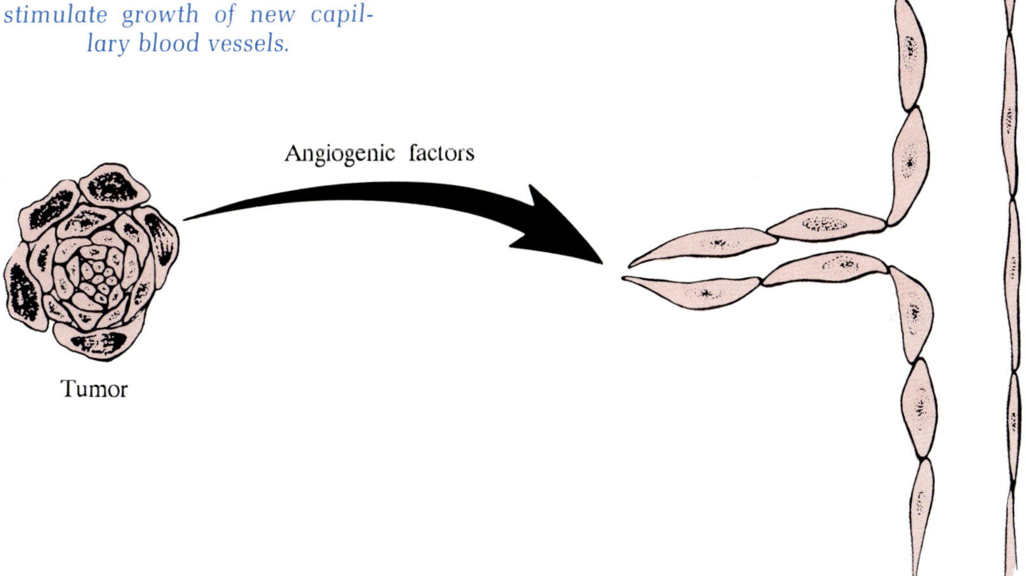

Figure 4. Tumors release diffusible angiogenic factors that travel short distances (millimeters) and stimulate growth of new capillary blood vessels.

In addition to its presence in many tumors, the angiogenic protein, FGF, has also been found in a variety of normal tissues, where it is present in relatively large quantity (e.g., from 50 μg/kg for most tissues up to 500 μg/kg in the pituitary). This situation raises an intriguing question. Why are the vascular endothelial cells of these normal tissues so quiescent? Why are they not continually proliferating in response to FGF, a potent angiogenic protein so widely distributed in normal tissue? As this question has been pursued experimentally, it has become clear that there may be several control mechanisms that, under normal conditions, guard vascular endothelial cells from exposure to potent endothelial growth factors.

One such guardian is the fact that FGF seems to be mainly sequestered within the cells that produce it. The FGF molecule itself appears to lack the information to be secreted from a cell by the conventional pathways that allow other proteins to be released. This fact may explain why there is essentially no detectable FGF in the blood. Whether or not tumors can turn on special mechanisms to secrete FGF is a question currently under intensive study in many laboratories. Some FGF does escape from normal cells, however, because one type of FGF, basic FGF, has recently been found to be stored in the basement membrane of certain normal tissues and blood vessels. In this form it remains inaccesible to vascular endothelial cells. If the tissue or its blood vessels are wounded, however, this angiogenic protein could possibly leak out and help to initiate new capillary growth. The enzymes produced by certain tumors that can lyse basement membrane may also act to release angiogenic protein from their storage site in the basement membrane. Thus, the capacity of a tumor to tap into a storehouse of angiogenic protein may be another back-up system for the recruitment of new capillary blood vessels that the tumor can use for its further growth. This speculation has yet to be supported experimentally.

How Angiogenesis Is Studied in the Laboratory

One of the difficulties in trying to understand the process of angiogenesis is that it cannot be studied in a test tube. New laboratory methods have had to be devised so that the process of capillary blood vessel growth could be observed directly. One very useful method has been the culture of chicken embryos in a dish. A chicken egg is cracked into a plastic Petri dish three days after fertilization (Figure 5). It is kept in an incubator at body temperature for three more days, at which time a thin vascular membrane, the chorioallantoic membrane, covers its surface. If a tumor or a tumor extract is placed on the membrane, the growth of new blood vessels can be observed over the next two or three days. This was one of the methods used to isolate and purify angiogenic proteins from tumors. Another method is to implant tumor cells into a rabbit eye, where the newly induced blood vessels can be observed directly. A third method uses pure capillary endothelial cells grown in vitro. Endothelial cell locomotion and proliferation are two of the major components of capillary growth in vivo and can be more precisely analyzed in these endothelial cultures.

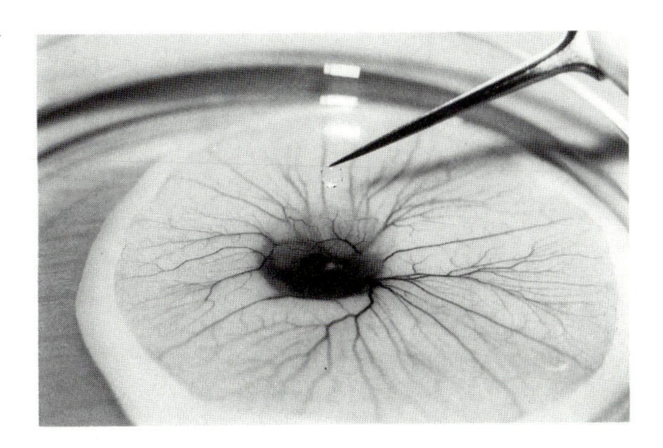

Figure 5. Chick embryo at six days of age growing in a Petri dish. A disk of methylcellulose (2-mm diameter) is about to be applied to the vascular membrane (chorioallantoic membrane) that is growing over the surface. This method is used to test angiogenic factors or angiogenesis inhibitors that can be dissolved or suspended in the methylcellulose vehicle.

How Capillary Blood Vessels Grow

Microscopic examination of blood vessels growing in the chick embryo and in other animals reveals that new capillaries are formed by a series of steps. After it is first stimulated by an angiogenic factor, an endothelial cell secretes enzymes to make a minute hole in the wall of a small vein (Figure 6). The hole is just large enough for a single endothelial cell to extrude itself outside of the blood vessel. Other endothelial cells then move in tandem through this break in the membrane and follow the leading endothelial cells toward the nest of tumor cells (Figure 7).

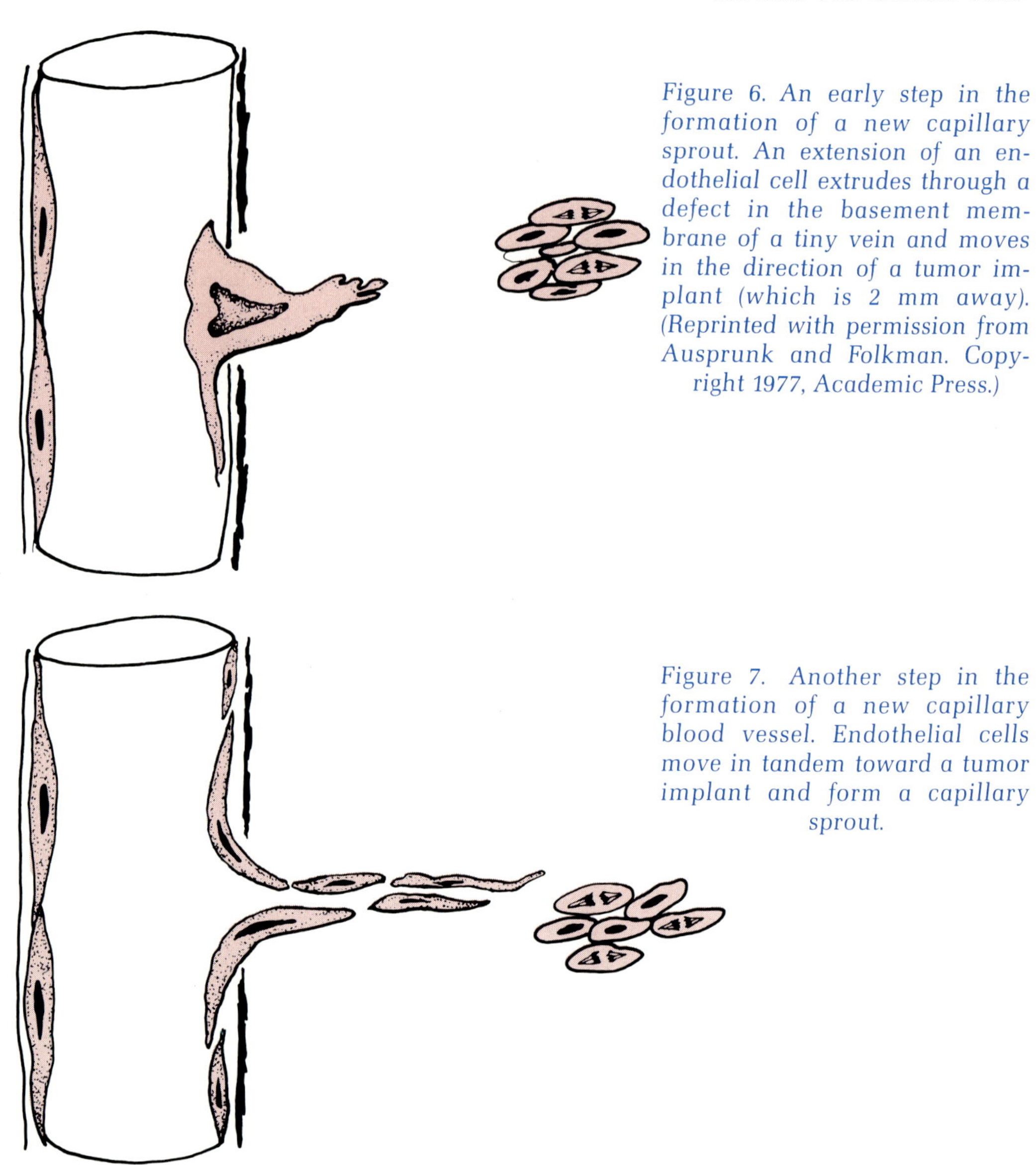

Figure 6. An early step in the formation of a new capillary sprout. An extension of an endothelial cell extrudes through a defect in the basement membrane of a tiny vein and moves in the direction of a tumor implant (which is 2 mm away). (Reprinted with permission from Ausprunk and Folkman. Copyright 1977, Academic Press.)

Figure 7. Another step in the formation of a new capillary blood vessel. Endothelial cells move in tandem toward a tumor implant and form a capillary sprout.

These endothelial cells form a sprout that soon becomes hollow as the endothelial cells curve themselves into a tubular configuration (Figure 8). The tip of each sprout then joins with another sprout to form a loop through which blood flows (Figure 9). New capillary sprouts arise from these loops, and the process is repeated until a large capillary network has developed. Tumor cells grow around each of these sprouts and loops. If one views a tumor approximately the size of a golf ball, these vessels are not easily visible. They can be seen, however, if after removal of the tumor the blood vessels are injected with a liquid plastic that is allowed to harden. When the tumor cells are removed by detergents or acid, the remaining vascular skeleton gives an accurate picture of how the tumor built itself up on the scaffold of new capillaries (Figure 10).

Figure 8. A capillary sprout elongating through collagen, as diagramed in Figure 7. The leading endothelial cell is in the lower right of the photograph. A hollow tube (lumen) is beginning to form in the midsection of this sprout. (Magnification approximately ×1400). (Reprinted with permission from DeVita et al. Copyright 1985, Lippincott)

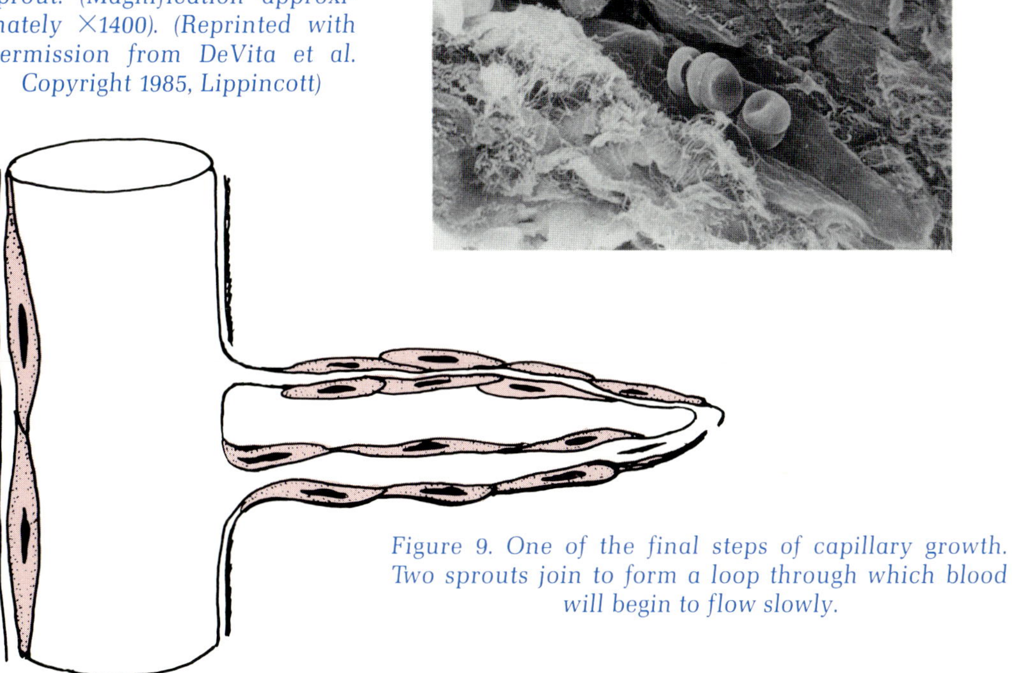

Figure 9. One of the final steps of capillary growth. Two sprouts join to form a loop through which blood will begin to flow slowly.

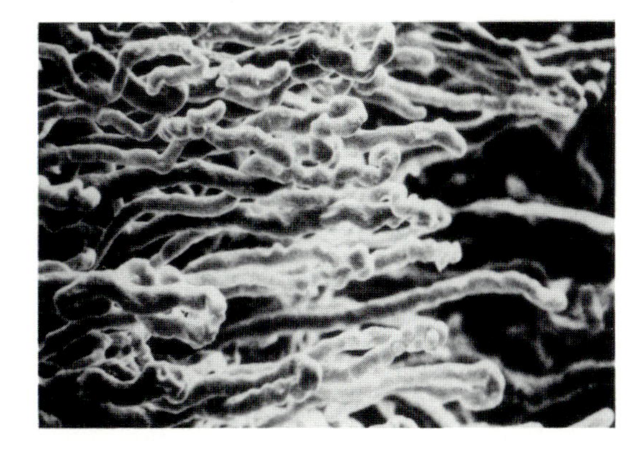

Figure 10. Capillary loops growing toward a human carcinoma of the larynx. The center of the tumor is on the right side of the photograph. The capillary loops are growing toward the right. In this specimen the tumor cells have been removed so that the blood vessels are more easily visualized (original magnification ×192). (Reprinted with permission from Miodonski et al. Copyright 1980, American Medical Association.)

Angiogenesis Inhibitors

After it was recognized both that tumor growth was angiogenesis-dependent and that tumors would remain static and tiny in the absence of new capillary blood vessels, scientists began to search for drugs that could prevent or stop capillary growth. These drugs, identified only in the past few years, are now known collectively as angiogenesis inhibitors. Currently, only about four angiogenesis inhibitors exist. It is quite possible that many more of these inhibitors will be discovered in the future.

One of the most potent angiogenesis inhibitors has been found in the blood itself. It is a breakdown product of cortisone, a hormone normally released into the blood from the adrenal gland. Cortisone has many functions, including the regulation of blood pressure and blood sugar, especially during stress or injury. The liver inactivates cortisone by breaking it down to a metabolite called tetrahydrocortisol, which is eventually excreted in the urine. Tetrahydrocortisol has none of the known activities of cortisone and was, until recently, thought to have no biological activity. We have recently discovered, however, that tetrahydrocortisol can inhibit capillary growth. It thus belongs to a new class of hormones called angiostatic steroids. This inhibitory activity is greatly enhanced if heparin is administered with the steroid. Heparin is commonly used to prevent clotting in patients.

Interestingly, nonanticoagulant fragments of the heparin molecule are also able to potentiate the ability of tetrahydrocortisol and certain other steroids to inhibit angiogenesis. Heparin and heparinlike molecules are also found normally on the surface of vascular endothelial cells and within the basement membrane of blood vessels and many other tissues. Recent evidence suggests that these heparinlike molecules may be acting together with circulating angiostatic steroids as natural suppressants of capillary growth. Higher doses of both drugs administered together can actually inhibit capillary growth and bring about regression of new capillary blood vessels. Angiostatic steroids have not yet been used to treat patients. However, they are a prototype of angiogenesis-inhibitor drugs that may someday be useful to treat certain tumors, as well as other nonneoplastic diseases dominated by abnormal angiogenesis.

Future Directions

As more is learned about the process of angiogenesis, it may someday be possible to diagnose early tumors as they become angiogenic. If we can understand how to block the release of angiogenic factors or how to inhibit their effect on blood vessels, it is conceivable that tumor growth may be controlled or at least made more susceptible to other kinds of anticancer chemotherapy.

The study of tumor angiogenesis has also helped to uncover the mechanisms of capillary blood vessel growth itself. This new knowledge may become useful in the understanding of certain nonneoplastic diseases that are dominated by abnormal angiogenesis, such as rheumatoid arthritis and diabetic retinopathy.

References

Ausprunk, D.; Folkman, J. *Microvascular Research* **1977,** *14,* 53–65.

Cady, B. *Cancer Manual*; American Cancer Society: New York, 1987.

DeVita, V. T.; Hellman, S.; Rosenberg, S. *Important Advances in Oncology 1985;* J. B. Lippincott: Philadelphia, 1985.

Jiang, L. W.; Schindler, M. *J. Cell Biol.* **1988,** *106* , 13–20.

Miodoński, A.; Kuś, J.; Olszewski, E.; Tyrankiewcz, R. *Arch. Otolaryngology* **1980,** *106,* 321–332.

Suggested Reading

Crum, R.; Szabo, S.; Folkman, J. "A New Class of Steroids Inhibits Angiogenesis in the Presence of Heparin or a Heparin Fragment." *Science (Washington, DC)* **1985,** *230,* 1375–1378.

Folkman, J. "The Vascularization of Tumors." *Scientific American* **1976,** *234,* 58–73.

Folkman, J.; Moscona, A. "The Role of Cell Shape in Growth Control." *Nature (London)* **1978,** *273,* 345–349.

Folkman, J.; Haudenschild, C. C.; Zetter, B. R. "Long-Term Culture of Capillary Endothelial Cells." *Proc. Natl. Acad. Sci. USA* **1979,** *76(10)*, 5217–5221.

Folkman, J. "Tumor Angiogenesis." In *Advances in Cancer Research;* Klein, G.; Weinhouse, S., Eds.; Academic: New York, 1985; 43, pp 175–203.

Folkman, J. "Toward An Understanding of Angiogenesis: Search and Discovery." In *Perspectives in Biology and Medicine.* The University of Chicago Press: Chicago, 1985; 29(1), pp 10–36.

Folkman, J. "Angiogenesis and Its Inhibitors." In *Important Advances in Oncology;* DeVita, V. T., Jr., Hellman, S.; Rosenberg, S.A., Eds.; J. B. Lippincott: Philadelphia, 1985; Part 1, pp 42–62.

Folkman, J.; Klagsbrun, M. "Angiogenic Factors." *Science (Washington, DC)* **1987,** *235*, 442–447.

Hudlicka, O.; Tyler, K. R. *Angiogensis: The Growth of the Vascular System;* Academic: New York, 1986.

Jiang, L.-W.; Schnidler, M. "Nuclear Transport in 3T3 Fibroblasts: Effects of Growth Factors, Transformation, and Cell Shape." *J. Cell Biol.* **1988,** *106*, 13–19.

Lapis, K.; Liotta, L. A.; Rabson, A. S. *Biochemistry and Molecular Genetics of Cancer Metastasis;* Martinus Nijhoff Publishers: Boston, 1986.

Peterson, H.-I. *Tumor Blood Circulation: Angiogenesis, Vascular Morphology in Blood Flow of Experimental and Human Tumors;* CRC Press: Boca Raton, 1979.

Rafkin, D. B.; Klagsbrun, M. *Angiogenesis. Mechanisms and Patholobiology in Current Communications in Molecular Biology;* Cold Spring Harbor Laboratory: New York, 1987.

CHAPTER **12** # Chemotherapy of Cancer

Joseph H. Burchenal and Joan Riley Burchenal

Chemotherapy has as its objective the use of chemicals to treat disease without seriously harming the patient. This objective is accomplished by killing a parasite as in malaria, killing a bacterium as in pneumonia, or preferentially destroying cancerous cells with a combination of or, rarely, a single anticancer drug.

The successful use of chemicals in treating systemic disease began when the Incas of Peru used a tea made from cinchona bark to combat malaria. The active component, quinine, was not isolated, however, until centuries later. In 1904, an important step forward in controlling infectious disease occurred when Paul Ehrlich (Ehrlich and Shiga, 1904) used laboratory mice to demonstrate that a dye, Trypan Red, could inhibit the growth of the parasite that causes sleeping sickness in horses, without harming the host. Then in 1910, after studying 605 compounds against the spirochete of syphilis, he finally found that the 606th, Salvarsan, would cure the early stages of this dread disease (Ehrlich, 1910). In 1935, Gerhard Domagk (Domagk, 1935) reported that Prontosil, the first of the sulfonamide drugs, could cure streptococcal infections. Chemically altered sulfonamides with a broader

1420–4/88/0201$06.50/0

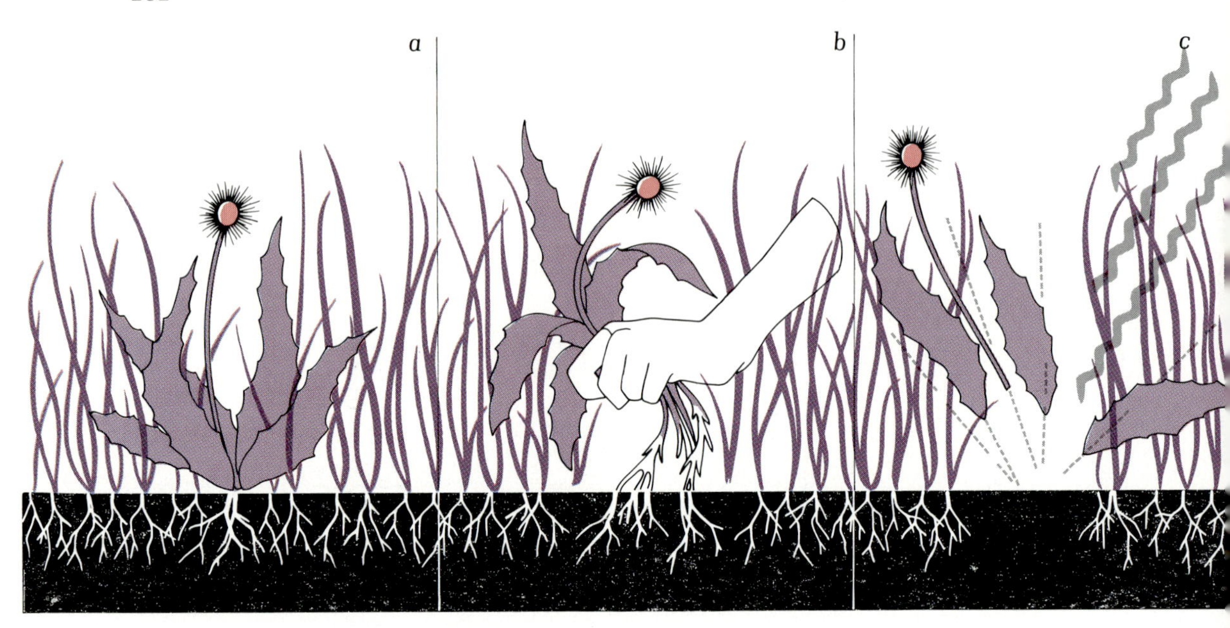

Figure 1. The usefulness of cancer treatment approaches can be illustrated using an analogy of weeds (cancer) and grass (normal cells). One weed (localized cancer) in the middle of a plot of grass can be pulled out (surgery) or burned out (radiotherapy). If the weeds are widespread (disseminated cancer), then a solution of weedkiller (chemotherapy) can be sprayed on the lawn to destroy the weeds selectively, without destroying the grass. If there are only a few weeds throughout the grass, the addition of a fertilizer (immunotherapy) to strengthen the growth of the grass may enable it to outgrow and choke out the weeds.

spectrum of activity soon followed. With the discovery (Fleming, 1929) and development in the 1940s (Chain et al., 1940) of penicillin, of streptomycin's action against tuberculosis, and of chloromycetin's activity against typhoid and typhus fevers, the era of the chemotherapy of infectious disease was launched. It appeared possible to use this technique at last against cancer.

Four Modes of Cancer Treatment

In addition to fighting infectious disease, chemotherapy has become one of the four major approaches used by physicians to destroy cancer cells selectively. Other approaches include surgery and radiotherapy for treatment of localized tumors, and immunotherapy, which aims to increase the patient's own resistance to the cancer.

An Analogy. The comparative usefulness of the four approaches can be illustrated by an analogy of weeds (cancer) and grass (normal cells). As illustrated in Figure 1, if there is one weed (localized tumor) in the middle of a plot of grass, it can be pulled out (surgery) or blasted out (radiotherapy). On the other hand, if the weeds are widespread

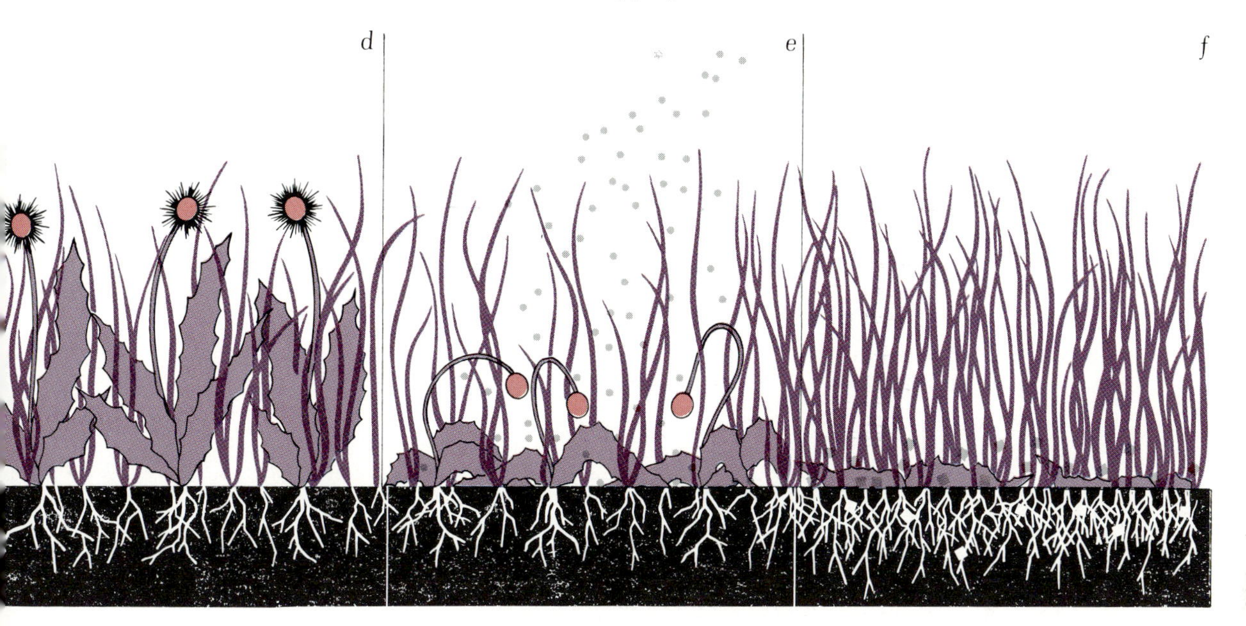

among the grass (disseminated cancer), then pulling them out or blasting them out would destroy the whole lawn. Instead, when a solution of weedkiller (chemotherapy) is sprayed on the lawn, the weeds are selectively destroyed without killing the grass. If there are only a few weeds throughout the grass, the addition of a fertilizer (immunotherapy) to strengthen the growth of the grass may make it outgrow and choke the weeds.

Treatment Limitations. The effectiveness of surgery is limited not by the size of the tumor but by its distribution. Even a very large tumor can be removed completely by surgery if it has not spread to another part of the body. Conversely, a small tumor that has dispersed even a few cells to other organs such as the lungs, liver, or brain cannot be treated successfully by removing the primary tumor alone. Similarly, radiotherapy in high doses can destroy a localized tumor, but these massive doses cannot be used over the whole body to destroy disseminated cells without seriously harming or killing the cancer patient.

Chemotherapy, on the other hand, is limited not by the spread of the disease but rather by its total mass. Although anticancer drugs penetrate the body and search out any clumps of cells that may have lodged in other organs, chemotherapy has great difficulty in destroying all cells in a large tumor. For example, a single course of therapy pushed to the limit of the patient's ability to tolerate it may destroy 99–99.9% of the tumor cells. In a large tumor weighing 1 kg and containing 10^{12} cells, a 99.9% kill would still leave a billion (10^9) living tumor cells. The same treatment, however, used against a relatively small clump of cells, 10^2 or 10^3, would leave essentially no living cancer cells. The limitation of disseminated tumor cells may be overcome in the future by the use of immunotherapy follow-up treatment to destroy the last few cells remaining after surgery, radiotherapy, or chemotherapy.

Unfortunately, chemotherapy destroys not only cancer cells, but also many rapidly dividing normal body cells such as cells lining the gastrointestinal tract, hair follicles, bone marrow cells, and lymphocytes involved in the immune defense system. This destruction of normal cells results in the common side effects of chemotherapy: nausea, vomiting, diarrhea, hair loss, and increased susceptibility to infection. The normal cells of the body, however, usually recover rapidly from these side effects after the course of treatment.

The Chemotherapeutic Approach

Chemotherapy's effectiveness depends on the nature of the cancer cell. In the process of becoming cancerous, the outlaw cell is no longer subject to the laws that control the growth of a normal cell. The cancer cell has changed some of its metabolic properties and may possess quantitatively different nutritional requirements or different enzymatic processes, compared to normal cells. In many respects, therefore, it acts in the same way as do foreign invaders such as bacteria, parasites, or viruses.

This being the case, many people ask why have cures been found for so many different types of infectious diseases and not for cancer. Although we do have many different drugs—some of which are active in pneumonia, some in typhoid fever, some in tuberculosis, and some in syphilis—we do not have a single chemical cure for all types of infectious disease. Similarly, in cancer, which is actually more than 100 different diseases, we clearly do not have a single cure for all cancers. However, we are able to cure some types of cancer. Although cancer cells differ somewhat from normal cells in their nutritional requirements, unfortunately they do not differ from other human cells as much as bacteria or fungi do. In fighting parasitic or bacterial diseases, qualitative differences in metabolism between the parasite and the host will often allow the infecting organism to be killed by drug levels that are harmless to the patient. Differences between the metabolism of cancer cells and the host cells, on the other hand, appear to be quantitative rather than qualitative, so that the amount of therapy given is limited by the tolerance of the patient's most sensitive cells. Because there are indeed small quantitative differences—nutritional requirements that differ in degree—it is possible for just the right dose of a compound to kill the cancer cells without irreversibly damaging the normal cells.

The logical approach, then, to finding new chemical agents active against cancer would be to find differences between the normal cell and the cancer cell that could be exploited in designing compounds to kill specifically the cancer cell. This rational approach, however, has not, so far, been of great clinical benefit. In general, the active drugs have been discovered on an empirical or observational basis. These compounds have demonstrated activity against tumors in tissue culture or in mice, and their actual mechanism of action has often been worked out years after discovery of their activity.

A certain amount of empiricism, however, has had a rational basis. When certain compounds have shown beneficial effects, many closely related derivatives have been synthesized and tested. Some of these derivatives have been great improvements over the original compounds in treating many types of cancer.

Anticancer Drugs. ***Antimetabolites.*** An antimetabolite is a compound not normally used by an organism but structurally very similar to an essential compound, much as two keys may have the same blank and therefore both fit in the same lock. The correct key has the notches in the correct places and is able to turn the lock. Although the wrong key, the antimetabolite, slides into the lock, the notches are in the wrong places, and it will not turn. But once in the lock, it prevents the normal key from getting in.

The first antimetabolite used in the effort to combat infectious disease was Prontosil Soluble, a complicated red dye found to have activity against streptococcal infection in mice. After many patients with streptococcal infections had been cured, it was discovered that only a very small portion of the molecule, the *p*-aminobenzenesulfonamide (sulfanilamide) moiety, was the active factor. The sulfanilamide acts as an antagonist, or antimetabolite: it prevents the streptococcus from using the *p*-aminobenzoic acid molecules necessary for its growth and survival. Because humans have no need for *p*-aminobenzoic acid, this is an example of the qualitative differences between the nutritional requirements of the streptococcal and human cells.

Use of the sulfonamides against infectious disease eventually led to the use of antimetabolites in the treatment of leukemia (Farber et al., 1948) and other types of cancer. The antimetabolites of folic acid, such as aminopterin and its improved derivative, methotrexate (Chart I), were the first to be of clinical value (Farber et al., 1948). These compounds are used to treat choriocarcinoma and several other types of cancer. After several courses of treatment against choriocarcinoma of the uterus, the tumor is entirely destroyed and the patient is once again completely healthy. In this particular case, the sensitivity of tumor cells to methotrexate is so great that a cure results. In acute childhood leukemia, however, although the right dose of aminopterin (Farber et al., 1948) or methotrexate will gradually destroy most of the leukemic cells and the patient will appear to go back to complete health for six months to a year or two, the administration of

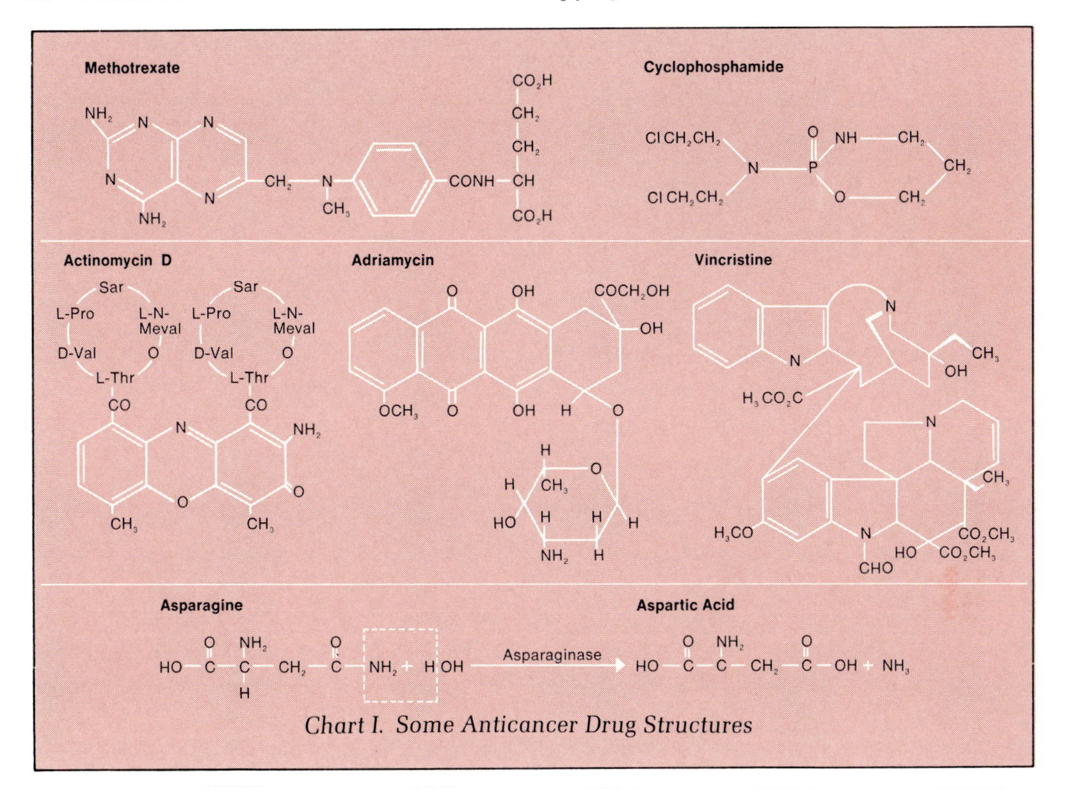

Chart I. Some Anticancer Drug Structures

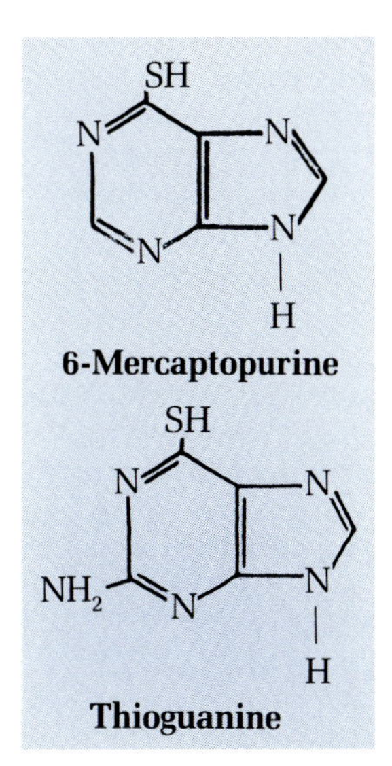

methotrexate alone is not usually sufficient to kill all the leukemic cells. The few cells remaining eventually develop a resistance to methotrexate and continue to grow, a situation causing a relapse despite continued methotrexate therapy. Fortunately, in these patients, a purine antagonist, 6-mercaptopurine (Elion et al., 1952; Burchenal et al., 1953), acting on a different metabolic pathway, is active even against methotrexate-resistant cells and even at daily doses low enough to allow the nonmalignant leukocytes, erythrocytes, and platelets to return to normal (Figure 2).

Another antimetabolite, 5-fluorouracil (5-FU) (Heidelberger, 1957), has been valuable in treating certain forms of gastrointestinal and breast cancer. In 5-FU, a fluorine atom is substituted for hydrogen in the 5 position of the pyrimidine ring of uracil. This carbon-fluorine bond is extremely stable and precludes the addition of a methyl group in the 5 position. The addition of a methyl group in the 5 position would result in the thymine analog of uracil—an important part of the DNA molecule.

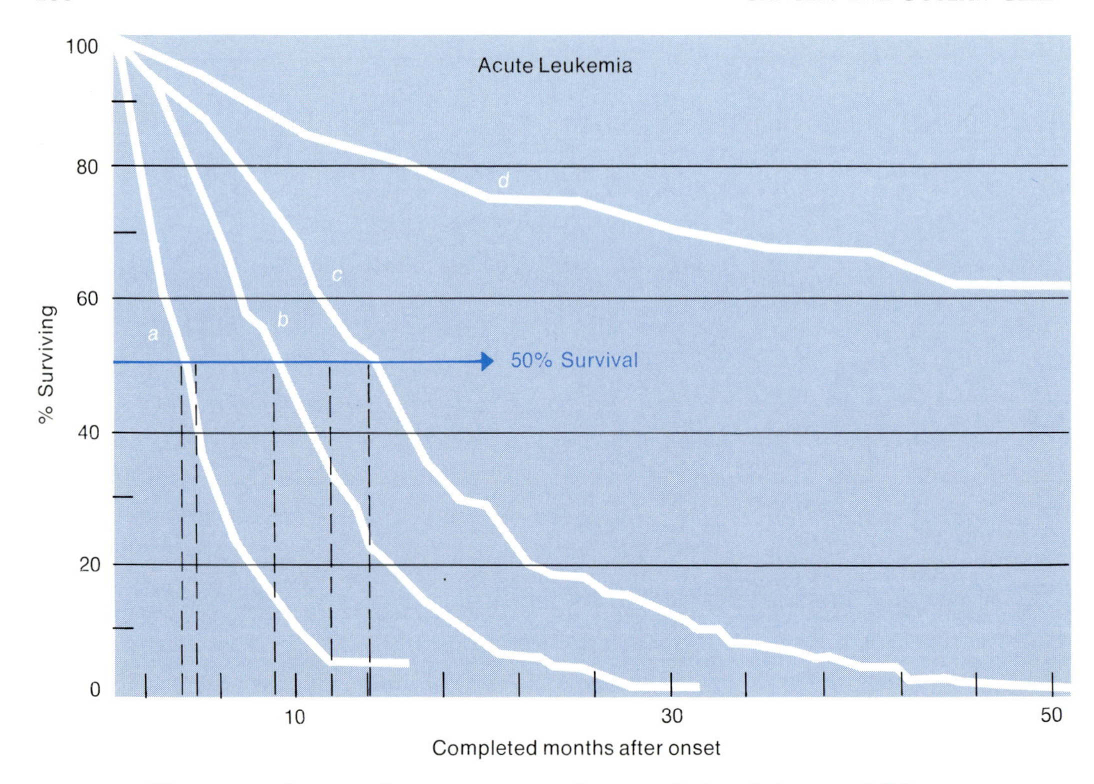

Figure 2. In certain cancers, great improvements in survival rates have been achieved with combined chemotherapy. (a) Forty years ago the median survival time for untreated children with acute lymphoblastic leukemia was between three and four months. (b) A report of 160 patients treated (Jan. 1948–Jan. 1952) with antimetabolites and steroids showed some increase in survival time. (c) Further improvement was demonstrated with treatment of 205 cases (Sept. 1959–April 1965) using steroids, cyclophosphamide, vincristine, and antimetabolites. (d) A later study involving 43 children (Nov. 1969–March 1972) used steroids, daunomycin, and vincristine, as well as two antimetabolites (methotrexate and 6-mercaptopurine) and one alkylating agent (cyclophosphamide). Of these patients, approximately 50% are free of disease and are presumably cured. More recent studies not shown in this figure (Clavell et al., 1986; Steinherz, et al., 1986; Lampert et al., 1984; Hammond, 1985), involving much larger groups of patients, have shown 60–90% five-year disease-free survival.

Thus, 5–FU interferes with the cancer cell's reproduction by interfering with the incorporation of thymidine into DNA. Arabinosylcytosine (Ara C), an antimetabolite of deoxycytidine, is of great value in protocols for treating acute nonlymphocytic leukemia in adults and children (Evans et al., 1964; Ellison et al., 1968).

Alkylating Agents. In addition to the antimetabolites, alkylating agents are particularly valuable in fighting cancer. Alkylating agents are very active compounds that biochemically alkylate important compounds such as DNA, RNA, and certain enzymes. Their original use stems from the observation that individuals heavily gassed with mustard gas, bis(β-chloroethyl) sulfide, during World War I suffered damage to their bone marrow and lymphoid tissues. Animal studies during World War II demonstrated that heavy exposure to the nitrogen mustards, bis(β-chloroethyl)amino compounds, destroyed lymphoid tissues. It was an obvious step to use these chemicals cautiously to treat patients with cancers of the lymphoid tissues, such as lymphosarcoma and Hodgkin's disease (Gilman et al., 1946). The drugs were highly successful in shrinking these tumor masses, but at the same time also damaged normal bone marrow. Eventually, the damage to the bone marrow during extended treatment made it impossible to continue therapy at high enough dosage to destroy the tumor completely.

Many derivatives of these alkylating agents were synthesized, with various improvements. One particularly interesting drug, cyclophosphamide (Brock and Wilmanns, 1958) (Chart I), was synthesized in the belief that it would be inactive in the body until its ring structure was broken down by an enzyme more common in cancer cells than in normal cells. It was thought that cyclophosphamide would be inert until it penetrated the cancer cell, where it would be converted to the active derivative and thus damage the cancer cell specifically. This theory did not turn out to be quite correct, however. Cyclophosphamide is converted to the active compound mainly in the liver, rather than in the tumor. Despite an incorrect premise concerning its mechanism of action, it seems superior to many of the other alkylating agents and is frequently used in treating lymphosarcomas and Hodgkin's disease, as well as breast, ovarian, and lung cancer.

Antibiotics. Antibiotics form yet another group of chemotherapeutic agents. The first of these, actinomycin D (Chart I), was discovered in the 1940s as an antibiotic against certain bacteria. It proved to be too toxic to the patient, however, to be of value in treating infectious disease. Ten years later, it was tried with considerable success in the treatment of various tumors, particularly Wilms' tumor (Farber et al., 1956), a kidney tumor in children, and also in a type of uterine cancer. Since then, thousands of antibiotics have been tested for antitumor activity. At present, three of the most successful antibiotic compounds are daunorubicin (Tan et al., 1967; Bernard, 1967), its closely related derivative adriamycin, and mitomycin. Adriamycin (Chart I) is of particular value in treating leukemias and lymphomas, as well as many other types of cancers. New semisynthetic derivatives of many of these agents, believed to be more active and less toxic, are under study.

Compounds from Natural Sources. A number of active agents are derived from natural plant sources. Two such compounds, vinblastine and vincristine (Chart I), are alkaloids derived from the periwinkle plant (*Vinca rosea*) (Noble et al., 1960; Warwick et al., 1960). These agents are particularly useful against leukemias and lymphomas, but they are also used against several solid tumors. Similarly, etopside and VM26, semisynthetic derivatives of podophyllin from the May apple (*Podophyllum peltatin*), are also useful in some solid tumors. Chemically modified derivatives of more of these agents from natural sources are now under clinical study.

Around 20 years ago there was great excitement about a new type of chemotherapy using the enzyme asparaginase originally found by Kidd in guinea pig serum (Kidd, 1953). This enzyme was quite active in treating lymphoma in mice and dogs, and it was thought that some cancer cells might be qualitatively different from normal cells in their asparagine-synthesizing enzyme system. Normal cells do not require an outside source of the amino acid asparagine because they have an enzyme, asparagine synthetase, that can synthesize asparagine from aspartic acid and glutamine. Certain leukemic and lymphoma cells, however, lack this enzyme. As a result, they depend completely on an outside source, such as the blood, for their supply of asparagine.

It was, therefore, felt that if one administered a large dose of asparaginase, an enzyme that removes the amino group from asparagine, all the asparagine circulating in the blood plasma would be destroyed, and the leukemic cells would be starved of this essential nutrient. Conversely, normal cells would continue to synthesize their own asparagine.

This therapy produced *complete remissions* in about 50% of the children with acute leukemia. Unfortunately, the leukemic cell eventually adapts to the altered situation by inducing the enzyme asparagine synthetase and then by synthesizing its own asparagine. After a few months, the patient's leukemic cells begin making adequate asparagine and become resistant to asparaginase therapy (Oettgen et al., 1967). Asparaginase, however, is still useful as part of the combination used in the induction phase of the therapy of acute leukemia in children.

Platinum and Other Heavy Metals. The astute discovery by Rosenberg et al. (1969) of the effects of a platinum coordinate complex, *cis*-diamminedichlorideplatinum (cisplatin), first on bacteria and then on mouse tumors and leukemias, led to clinical trial and the demonstration of activity in patients with various types of tumors, particularly testicular and ovarian cancers. Cisplatin is nephrotoxic and neurotoxic, but its nephrotoxicity may be largely prevented by the administration of large amounts of intravenous fluids. This method allows higher doses to be given. Newer derivatives, such as carboplatin, are less nephrotoxic and neurotoxic, and are limited at higher doses only by a reversible bone marrow depression. Several 1,2-diaminocyclohexane (DAC) platinum compounds are active in mice against tumors and leukemias that have developed a high degree of resistance to Cisplatin. Compounds of this series have not received adequate trial in patients because of solubility and stability problems, but a new DAC platinum(IV) derivative, tetraplatin (Engineer et al., 1987), is now in preclinical pharmacology and appears to have considerable promise. Platinum compounds are useful in combination therapy of testicular (Einhorn, 1981), ovarian, head and neck, and lung tumors. Coordinate complexes of gold and of palladium are also being studied in animal tumors and leukemias, but so far without great success.

Combination Chemotherapy. The development of cells resistant to any single agent is a major problem in using chemotherapeutic agents, and combination chemotherapy has been developed to avoid resistance. There is usually no cross-resistance between different types of agents with different mechanisms of action. For example, a leukemic cell that develops resistance to mercaptopurine may still be responsive to methotrexate, cytoxan, vincristine, or adriamycin, and vice versa. Therefore, in combination chemotherapy, the few cancer cells resistant to one of the chemicals can still be killed by the other drugs, because the chances of a cell's developing resistance simultaneously to all drugs in a combination are very small.

Even with the relatively small number of different drug types currently available, intensive combination chemotherapy—using five or six different agents—has greatly improved the outlook of patients with acute leukemia. Forty years ago the median survival time for children with acute lymphoblastic leukemia was between three and four months. The results of therapy have been improving steadily (Figure 2). Now 60–90% of the patients can be expected to be free of the disease at five years, and most of these will go on to be cured (Hammond, 1985). One very important step toward this goal has been recognizing certain factors, such as high leukocyte count, as predictors for poor response to standard therapy. For patients in this group, even more intensive regimens of treatment have greatly improved long-term, disease-free survival (Lampert et al., 1984; Clavell et al., 1986; Steinherz et al., 1986).

Following the discovery of the curative potential of methotrexate and cyclophosphamide in Burkett's tumor (Oettgen et al., 1963), a fulminating lymphoma of African children, intensive combination therapy by Wollner et al. began to produce cures in a high percentage of widespread lymphosarcoma and Hodgkin's disease in children (Wollner et al., 1973). Results have not been quite as satisfactory in acute myeloblastic leukemia of adults, because there are fewer active agents and they are less specific. Complete remissions in 50–80% of these patients can be achieved, and perhaps 15–25% will be free of disease after three to five years.

Bone Marrow Transplant. A newer, more radical, and apparently more successful approach to leukemia is used in the treatment of children and adults under the age of 40 (Thomas, 1983). At the time of the first remission with acute myeloblastic leukemia or second remission in acute lymphoblastic leukemia, the patient is offered a bone marrow transplant if he or she has a sibling with an identical tissue type who can donate marrow. The patient is then given whole-body radiation either as a single dose of 1000 rads or as a total of 1200–1320 rads divided into 120 rads every eight hours for 10–11 doses. In addition, a dose of 60 mg/kg of cyclophosphamide on each of two consecutive days is given either immediately before or after radiation. This combination would be lethal without marrow transplant and is calculated to destroy the few leukemic cells presumably remaining during a complete remission.

After 24 to 96 hours, bone marrow aspirated from the iliac bones of the donor is infused intravenously. These cells home in on the now-depleted marrow of the patient and usually engraft successfully. Thus the patient's leukocyte and platelet counts are restored with cells derived from the donor marrow. For a temporary period, the patient's defenses against infections are hazardously low and protection with antibiotics must be given, but if engraftment occurs and graft-versus-host reactions due to minor incompatibilities between donor and host are not too severe, most patients should return to normal health and be free of leukemia indefinitely. Approximately 50% complete cures are achieved by this somewhat complicated and hazardous technique (Thomas, 1983; Brochstein et al., 1987).

In advanced Hodgkin's disease of adults—previously considered a sentence of death—at least 50% of patients treated with the MOPP program (HN2, vincristine, procarbazine, and prednisone) of DeVita et al. (1976) are free of disease at five years, and 50% of those relapsing can be cured with further courses of a different combination ABVD (adriamycin, bleomycin, vinblastine, and imidazole-carboxamide) developed by Bonadonna (Bonadonna et al., 1974). Even better results are now being achieved in some treatment series by alternating the two regimens. Complete, long-lasting, and presumably permanent remissions have been reported in 90% of young men with metastatic testicular cancer (Einhorn, 1981), again by strenuous combination therapy.

Multidisciplinary Therapy. Cancers of the blood or blood-forming organs have been shown to be relatively sensitive to chemotherapy. In treating cancers of the bones and the more common solid cancers of the lung, breast, and colon, the roles of the other forms of cancer treatment, such as surgery and radiation, and their relation to chemotherapy must be considered.

As mentioned earlier, chemotherapy is used in treating any cancer that has spread beyond the ability of surgery or radiotherapy to control it. But most frequently when the cancers have reappeared after surgery, they are present in large amounts, and the drugs or combination of drugs used are not sufficiently active to destroy all the cancer cells. If the number of tumor cells is small at the start of chemotherapy, there is a much better chance that there will be no viable cells remaining after treatment. When the surgery and radiation—alone or in combination—are used to destroy the bulk of the tumor and when chemotherapy is given immediately to destroy any remaining cancer cells, certain high-risk cancers can be controlled successfully. For example, this adjuvant chemotherapeutic approach has been used in patients with osteogenic sarcoma, a bone cancer in young adults frequently occurring in the extremities. Complete removal of the local tumor by amputating the leg successfully cures only about 20% of the patients, because within 6–18 months after surgery, tumor growth will appear in the lungs of the other 80%. Some cancerous metastases must have been present in the lung at the time of surgery, but were so small that they could not be seen on the X-ray film. If, however, intensive chemotherapy is given (Jaffe et al., 1972; Cortes et al., 1972; Rosen et al., 1981) before and immediately after surgery, these small deposits of cells will be destroyed, and 70–80% of the patients will survive free of cancer. A few patients still develop pulmonary metastases, but these can be removed by thoracic surgery and chemotherapy.

This type of multidisciplinary therapy is also applied to certain other high-risk cancers, such as cancer of the breast that has spread to the axillary lymph nodes under the arm and cancer of the large bowel that has spread to the lymph nodes in the bowel wall. Striking early results have been reported by Bonadonna (Bonadonna et al., 1976). A total of 386 patients with breast cancer and axillary node involvement were assigned randomly to one of two treatment groups—radical surgery with no chemotherapy or radical surgery followed by a two-week course each month of cyclophosphamide, methotrexate, and 5-FU (CMF). In the preliminary report, there was a 24% recurrence among those patients who were treated by radical mastectomy alone. However, in those patients treated with radical mastectomy plus adjuvant chemotherapy, the recurrence rate was 5.3%. These results are for the two groups overall. They do not take into consideration differences among the postmenopausal and premenopausal groups, the extent of nodal involvement, or other forms of treatment administered earlier. The general consensus now is that in premenopausal women with three or less axillary nodes, the therapeutic combination of cytoxan, methotrexate, and fluorouracil (CMF) or cytoxan, adriamycin, and fluorouracil (CAF) are the regimens of choice and are definitely able to increase long-term survival. In premenopausal women with more than three positive axillary nodes, the benefit is less clear-cut. In postmenopausal women with estrogen receptor (ER) positive tumors, the addition of tamoxifen also appears to be beneficial.

These results suggest that adjuvant chemotherapy may increase the cure rate in certain types of patients who have a high risk of recurrence following surgery or radiation therapy alone. If this hypothesis is correct, as seems indicated now, it will give tremendous impetus to the attack on the other big cancer killers—cancer of the lung, pancreas, and colon. Better combinations of treatment methods including two or more disciplines among surgery, radiotherapy, chemotherapy, and immunotherapy will be used to the best advantage in attempting to eradicate cancer.

New Approaches

Screening. In hopes of improving the yield of active compounds from the Cancer Chemotherapy National Service Center's (CCNSC) Screening Program, particularly those that might prove useful against specific human solid tumors, a primary in vitro screen has been implemented in the past year using several cell lines derived from each of the most common human cancers, such as small and nonsmall cell cancers of the lung, breast, colon, and ovary. In present clinical oncology some agents are active against one tumor but not against others. Particular attention would be paid to compounds active, for instance, against lung but not against breast or colon, or active against colon but not against lung or breast. Such a compound would then be tried, where possible, against the same tumor growing in vivo in nude mice. If it inhibited the growth of the tumor or destroyed it entirely at tolerated doses, it was hoped that such specific antitumor activity would be extrapolated to patients in clinical trials. Unfortunately, not all of the in vitro tumor lines will grow in the nude mouse. No new clinically active compounds have so far been uncovered by this combined technique, but at least some have been positive at the in vitro stage and are awaiting trial in mice.

The rationale behind this system is good, but in the last analysis it is necessary to "test the test." Over 20 years ago, a large in vitro screening program was developed by the CCNSC. It produced several compounds that were highly active at miniscule concentrations against tumor cell lines in vitro. Unfortunately, these were equally toxic, or more so, against other organ systems in both mouse and humans, and thus had no therapeutic value. It is hoped that the present system in which compounds active in vitro are then tested for activity against the same human tumor growing in the nude mouse will demonstrate the specific activity of the agent (Chabner, 1986). Unfortunately, because the program is still undergoing minor modifications, there are no exact descriptions or references to this other than the excellent article in the *New York Times* Science Section of Tuesday, December 23, 1986.

Mechanisms of Resistance. The problem with all clinically active agents is the eventual development of resistance. Resistance to methotrexate was first developed in the AK4 line of mouse leukemia in 1951 and, later that year, in L1210. Since then the problem of chemotherapeutic resistance has been studied extensively in vitro and in vivo. Resistance is thought, in most cases, to be a random mutation, which is selected out by the fact that the mutated cell is better able to cope with what would otherwise be toxic concentrations of the compound. Thus, several different mechanisms may be causing resistance to a given agent. For example, with methotrexate, resistance may be due to greatly increased concentrations of dihydrofolate reductase (DHFR), the enzyme inhibited by MTX with selective amplification of the DHFR gene, or to the production of an altered DHFR, or to changes in the transport mechanism, or to defective polyglutamylization. Correlation of karyotypic changes with some types of resistance is well documented (Albrecht and Biedler, 1984).

With other agents different mechanisms of resistance may be causative. Of particular interest are the pleotropic (multiple drug) resistant (MDR) mutants, in which, with lines resistant to adriamycin, for instance, there may be cross-resistance to vincristine, etopside, VM26, and dactinomycin (Chabner, 1986). Generally, this MDR cross-resistance occurs with the products derived from natural sources, but it may occasionally extend to other agents such as cisplatin. At other times cross-resistance may be limited to compounds of related molecular structure.

Autologous Marrow Transplant. Recently, autologous marrow transplant has been useful in some patients with widespread solid tumors, but without bone marrow involvement with tumor cells. Autologous marrow transplant contrasts with the allogeneic transplantation discussed previously: in the newer procedure a portion of the patient's own pelvic marrow is removed by aspiration and frozen until required. The patient is then given massive doses of drugs whose primary or only toxicity is myelosuppression (marrow suppression). Several such drugs—nitrogen mustard, cyclophosphamide, triethylenephosphoramide, phenylalanine mustard, etopside, carboplatin, vinblastine—have no significant cross-resistance. Thus an enormous cumulative antitumor effect can be exerted with toxicity only to the marrow, which will be destroyed. A few days later the patient's own stored and untreated marrow is returned to replace the marrow destroyed by the cumulative combined effects of the treatment. This repopulates the marrow cavity and normal marrow function returns, although the tumor is either completely or mostly destroyed. It is too early to evaluate the success of this technique, but the rationale is good and early studies are promising (Antman et al., 1987).

Molecular Biology. Studies on the molecular biology of cancer and on various biological modifiers have made great strides in the past few years. Interferon has shown beneficial effects in patients with two types of leukemia—hairy cell leukemia and the chronic myelocytic leukemia—and is now produced by recombinant DNA methods. In addition, interleukin II and granulocyte and granulocyte macrophage colony stimulating factors (G-CSF and GM-CSF) have had their gene structures decoded and are being produced and studied in patients with promising results, particularly in counteracting the marrow depression produced by some chemotherapy. Thus, the years of probing into the molecular biology of cells, both normal and cancerous, have begun to pay dividends in the form of clinically useful substances. It appears likely that the instances described represent only the tip of the iceberg, and many more substantial advances will occur in the future.

Summary

There is no situation in medicine where close collaboration between the laboratory bench and the patient's bedside is more important than in the approach to the cancer problem. Here synthetic organic chemists and biochemists working closely with pharmacologists and clinicians may be able to achieve together what no discipline can achieve alone—the complete destruction of all the cancer cells and the cure of the cancer patient.

References

Albrecht, A. M.; Biedler, J. L. "Acquired Resistance of Tumor Cells to Folate Antagonists." In *Folate Antagonists As Therapeutic Agents;* Sirotnak, F. M.; Burchall, J. J.; Ensminger, W. B. Eds. Academic: New York, 1984; Vol. 1, pp 317–353.

Antman, K.; Eder, J. P.; Frei, E., III. "High-Dose Chemotherapy with Bone Marrow Support for Solid Tumors. In *Important Advances in Oncology;* DeVita, V. T., Jr.; Hellman, S.; Rosenberg, S. A., Eds.; B. Lippincott: Philadelphia, 1987; Vol. 3, pp 221–235.

Bernard, J. "Acute Leukemia Treatment." *Cancer Res.* **1967,** *27,* 2565.

Bonadonna, G., et al. "Combination Chemotherapy as an Adjuvent Treatment in Operable Breast Cancer." *N. Engl. J. Med.* **1976,** *294,* 405.

Bonadonna, G.; DeLena, M.; Uslenghi, C. et al. "Combination Chemotherapy of Advanced Hodgkin's Disease (HD): Adriamycin (ADM), Bleomycin (BLM), Vinblastine (VLB), and Imidazol Carboxamide (DTIC) Versus MOPP." *Proc. Am. Assoc. Cancer Res.* **1974,** *15,* 90.

Brochstein, J. A.; Kernan, N. A.; Groshen, S., et. al. "Allogeneic Bone Marrow Transplantation After Hyperfractionated Total-Body Irradiation and Cyclophosphamide in Children with Acute Leukemia." *N. Engl. J. Med.* **1987,** *317,* 1618.

Brock, N.; Wilmanns, H. "Wirkung eines zyklischen N-Lost-Phosphamidesters auf experimentell erzeugte Tumoren der Ratte; Chemotherapeutische Wirksamheit und pharmakologische Eigensch-'ten von B518 ASTA." *Dtsch. Med. Wochenschr.* **1958,** *83,* 453.

Burchenal, J. H.; Murphy, M. L.; Ellison, R. R., et al. "Clinical Evaluation of a New Antimetabolite, 6-Mercaptopurine, in the Treatment of Leukemia and Allied Diseases." *Blood* **1953,** *8,* 965.

Chabner, B. A. Karnofsky Memorial Lecture. *J. Clin. Oncol.* **1986,** *4,* 625.

Chain, E.; Florey, H. W.; Gardner, A. D., et al. "Penicillin as a Chemotherapeutic Agent." *Lancet* **1940,** *2,* 226.

Clavell, L. A.; Gelber, R. D.; Cohen, H. J., et al. "Four-Agent Induction and Intensive Asparaginase Therapy for Treatment of Childhood Acute Lymphoblastic Leukemia." *N. Engl. J. Med.* **1986,** *315,* 657.

Cortes, E. P.; Holland, J. F.; Wang, J. J., et al. "Doxorubicin in Disseminated Osteosarcoma." *J. Am. Med. Assoc.* **1972,** *221,* 1132.

DeVita, V.; Canellos, G.; Hubbard, S., et al. "Chemotherapy of Hodgkin's Disease (HD) with MOPP: A 10-Year Progress Report." *Proc. Am. Assoc. Clin. Oncol.* **1976,** *17,* 269.

Domagk, G. "Ein Beitrag zur Chemotherapie der bakteriellen Infektion." *Dtsch. Med. Wochenschr.* **1935,** *61,* 250.

Ehrlich, P. "Die Behandlung der Syphilis mit dem Ehrlichschen Praparat 606." *Vers. Ges Dtsch. Naturf Arzt.* **1910,** *82.*

Ehrlich, P.; Shiga, K. "Farbentherapeutische Versuche bei Trypanosomenerkrankung." *Berl. Klin. Wochenschr.* **1904,** *41,* 329.

Einhorn, L. H. "Testicular Cancer as a Model for a Curable Neoplasm." The Richard and Hinda Rosenthal Foundation Award Lecture. *Cancer Res.* **1981,** *41,* 3275–3280.

Elion, G. B.; Burgi, E.; Hitchings, G. H. "Studies on Condensed Pyrimidine Systems. IX. The Synthesis of Some 6-Substituted Purines." *J. Am. Chem. Soc.* **1952,** *74,* 411.

Ellison, R. R.; Holland, J. F.; Weil, M., et al. "Arabinosylcytosine, A Useful Agent in the Treatment of Acute Leukemia in Adults." *Blood* **1969,** *32,* 507.

Engineer, M. S.; Ho, D. W. H.; Brown, N. S., et al. "Comparison of the Nephrotoxicity of Tetraplatin and Cisplatin in Rats." *Proc. Am. Assoc. Cancer Res.* **1987,** *28,* 444.

Evans, J. S.; Musser, E. A.; Bostwick, L., et al. "The Effect of 1-β-D-Arabinosylcytosine on Murine Neoplasms." *Cancer Res.* **1964,** *24,* 1285.

Farber, S.; Diamond, L. K.; Mercer, R. D., et al. "Temporary Remissions in Acute Leukemia in Children Produced by the Folic Acid Antagonist, 4-Aminopteroylglutamic Acid (Aminopterin)." *N. Engl. J. Med.* **1948,** *238,* 787.

Farber, S.; Maddock, C.; Swaffield, M. "Studies on the Carcinolytic and Other Biological Activity of Actinomycin D." *Proc. Am. Assoc. Cancer Res.* **1956,** *2,* 104.

Fleming, A. "On the Antibacterial Action of Cultures of a Penicillium with Special Reference to Their Use in the Isolation of B. Influenza." *Br. J. Exp. Pathol.* **1929,** *10,* 226.

Gilman, A.; Goodman, L.; Lindskog, G. E.; et al. Classified Research 1942–1943 quoted in Gilman, A.; Philips, F. S. "The Biological Actions and Therapeutic Applications of the β-Chloroethyl Amines and Sulfides." *Science (Washington, DC)* **1946,** *103,* 409.

Hammond, G. D. "Multidisciplinary Clinical Investigation of the Cancers of Children. A Model for the Management of Adults with Cancer." Lucy Wortham James Lecture. *Cancer* **1985,** *55,* 1215.

Heidelberger, C.; Chanduri, N. K.; Danneberg, P. "Fluorinated Pyrimidines, A New Class of Tumor-Inhibitory Compounds." *Nature (London)* **1957,** *179,* 663.

Jaffe, N., et al. "Recent Advances in the Chemotherapy of Osteogenic Sarcoma." *Cancer* **1972,** *30,* 1627.

Kidd, J. G. "Regression of Transplanted Lymphomas Induced *in vivo* by Means of Normal Guinea Pig Serum." *J. Exp. Med.* **1953,** *98,* 565.

Lampert, F.; Henze, G.; Langermann, H. J., et al. "Acute Lymphoblastic Leukemia: Current Status of Therapy in Children." In *Recent Results in Cancer Research,* Vol. 93, Springer-Verlag: Berlin-Heidelberg, 1984.

Noble, R. L.; Beer, C. T.; Cutts, J. H. "Role of Chance Observations on the Effects of Vincaleukoblastine with Special Reference to Hodgkin's Disease." *Proc. Canadian Can. Res. Conf.,* **1960,** p 373.

Oettgen, H. F.; Burkitt, D.; Burchenal, J. H. "Malignant Lymphoma Involving the Jaw in African Children: Treatment with Methotrexate." *Cancer* **1963,** *16,* 616.

Oettgen, H. F.; Old, L. J.; Boyse, E. A., et al. "Inhibition of Leukemias in Man by L-Asparaginase." *Cancer Res.* **1967,** *27,* 2619.

Rosen, G.; Nirenberg, A.; Caparros, B., et al. "Osteogenic Sarcoma: Eighty-Percent, Three-Year, Disease-Free Survival with Combination Chemotherapy (T-7)." *Natl. Cancer Inst. Monogr.* **1981,** *56,* 213.

Rosenberg, B.; Van Camp, L.; Trosko, J. E., et al. "Platinum Compounds. New Class of Potent Antitumor Agents." *Nature (London)* **1969,** *222,* 385.

Steinherz, P. G.; Gaynon, P.; Miller, D. R., et al. "Improved Disease-Free Survival of Children with Acute Lymphoblastic Leukemia at High Risk for Early Relapse with the New York Regimen - A New Intensity Therapy Protocol: A Report from the Children's Cancer Study Group." *J. Clin. Oncol.* **1986,** *4,* 744.

Tan, C.; Tosaka, H.; You, K. P., et al. "Daunomycin, an Antitumor Antigiotic in the Treatment of Neoplastic Disease." *Cancer* **1967,** *20,* 333.

Thomas, E. D.; "Marrow Transplantation for Malignant Diseases." *J. Clin. Oncol.* **1983,** *1,* 517.

Warwick, O. H.; Darte, J. M. M.; Brown, T. C. "Some Biological Effects of Vincaleukoblastine, an Alkaloid in *Vinca rosea Linn* in Patients with Malignant Disease." *Cancer Res.* **1960,** *20,* 1032.

Wollner, N.; D'Angio, G. J.; Burchenal, J. H., et al. "Treatment of Non-Hodgkin's Lymphoma in Children with Multiple Drug Leukemia Regimen and Radiation Therapy." *Proc. Am. Assoc. Cancer Res.* **1973,** *14,* 97.

CHAPTER 13 Radiation Therapy

William D. Powlis, Philip Rubin, and Diana Furst Nelson

Three standard approaches to the treatment of cancer are radiation therapy, surgery, and chemotherapy. Chemotherapy is used primarily to treat disease that has metastasized; radiation therapy and surgery are used alone or in combination to treat localized tumors. Radiation therapy is superior to surgery when it effectively destroys a tumor while causing only minimal damage to the surrounding normal tissue. For example, in treating localized cancer of the larynx, surgery would require its complete removal, whereas radiation therapy allows the larynx to remain intact while achieving comparable survival rates. Radiotherapy of localized breast cancer has evolved into a favored treatment because local control and survival rates are comparable to surgical results. Instead of removing the whole breast (mastectomy), excision of the cancer (lumpectomy) is performed and the remaining breast is irradiated; the result is a cosmetically superior appearance.

Combining surgery and radiotherapy often improves the local control of a tumor. Radiotherapy can destroy microscopic cancer cells that may remain in the lymph nodes and surrounding tissues following surgery. Moreover, radiotherapy before surgery can reduce large tumors, decrease local recurrence, and may decrease the chance of metastasis. In addition, radiotherapy can be combined

1420–4/88/0223$06.00/0

with chemotherapy and immunotherapy to more effectively treat cancer patients whose disease has spread, but who also have a substantial risk of developing recurrent cancer in the site or sites of initial involvement.

Types of Radiation

Radiotherapy involves irradiation that is controlled, both in the amount absorbed and in the specific target. X-rays generated by specially designed machines (linear accelerators) and gamma rays from naturally occurring radioactive isotopes are used most often in radiotherapy against cancer. X-rays and gamma rays are emitted in the form of quanta or photons, each having an electromagnetic wave with a specific frequency and wave length.

As X-rays and gamma rays travel through the body's tissues, they interact with electrons. With low- or moderate-energy radiotherapy machines, the energy transmitted to tissues decreases as radiation travels deeper into the body. This means that healthy tissue nearer the skin's surface receives a larger dosage of radiation than does a deep tumor lying below it. Therefore, the tolerance level of healthy superficial tissue places a limit on the radiation dosage used to destroy a tumor.

With modern higher-energy X-rays, larger doses can be administered without exceeding the skin's tolerance. In fact, the maximum tumor dose that can be delivered by higher-energy X-rays is not limited by the skin dose, but by the amount tolerated by normal tissue surrounding the tumor. Late radiation effects, which may not be observed for months or even years after treatment, restrict the actual dosage that can be delivered and absorbed by normal tissue without risk of seriously impairing organ function.

The radiation dose that impairs organ function varies with different organs and tissues. Functional impairment can occur at doses of about 2000 rad in certain organs, for example, bone marrow, lung, kidney, or actively growing bone. Other tissues and organs, such as muscle and cutaneous tissue, do not manifest irreparable injury even at doses of 6000 rad or more. (A rad is a unit of measure of the amount of radiation absorbed by the body's tissues.) Tolerance, the ability to accumulate and repair radiation injury, is determined by a complex intermix of multiple factors including the volume irradiated, the proportion of an organ treated, the number of treatments (fractions) used to deliver a total dose, the dose rate (the time to deliver a particular dose, minutes vs. hours), the overall time to reach a total dose, the level of oxygenation, the type of radiation used, and so on. Because of this complexity, tolerances of specific organs have been learned empirically. Vital organs with the lowest radiation tolerance include the lungs, kidneys, liver, and spinal cord. These vital organs can be shielded during radiotherapy so that a tumor-destroying dose of radiation can often be delivered without damaging them.

Recent work in radiotherapy involves the use of densely ionizing radiations. Unlike conventional radiation, densely ionizing radiations such as protons deposit most of their energy over a short distance and produce a higher ionization density than do X-rays. As ionization density increases, tissue damage also increases. In addition, these heavy charged particles allow the radiation damage to be concentrated in the tumor itself. When the particles reach the tumor, the energy deposited rises sharply to a maximum. Thus a high dose can be delivered to the tumor and a low dose to the normal tissue beyond and surrounding the tumor. Although still experimental, densely ionizing radiations hold great promise for more effective radiotherapy.

Mechanism of Action

Radiation destroys cancer cells by damaging the DNA, causing biochemical changes that interfere primarily with the cell's ability to divide indefinitely. This process involves mechanisms on many levels—physical, chemical, biochemical, cellular, and tissue (Figure 1).

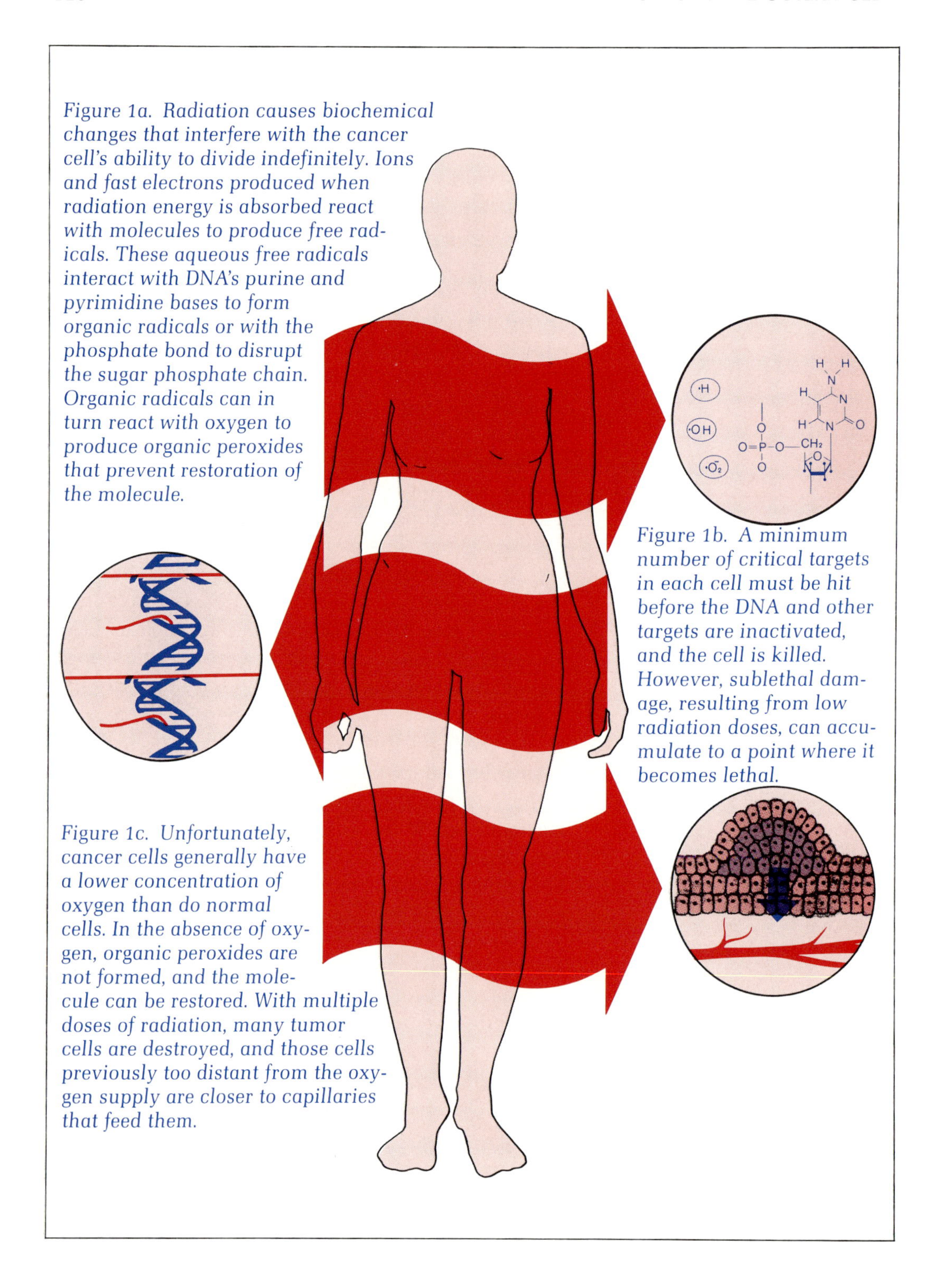

Figure 1a. Radiation causes biochemical changes that interfere with the cancer cell's ability to divide indefinitely. Ions and fast electrons produced when radiation energy is absorbed react with molecules to produce free radicals. These aqueous free radicals interact with DNA's purine and pyrimidine bases to form organic radicals or with the phosphate bond to disrupt the sugar phosphate chain. Organic radicals can in turn react with oxygen to produce organic peroxides that prevent restoration of the molecule.

Figure 1b. A minimum number of critical targets in each cell must be hit before the DNA and other targets are inactivated, and the cell is killed. However, sublethal damage, resulting from low radiation doses, can accumulate to a point where it becomes lethal.

Figure 1c. Unfortunately, cancer cells generally have a lower concentration of oxygen than do normal cells. In the absence of oxygen, organic peroxides are not formed, and the molecule can be restored. With multiple doses of radiation, many tumor cells are destroyed, and those cells previously too distant from the oxygen supply are closer to capillaries that feed them.

Physical and Chemical. Energy is transferred from the incident photon to the surrounding matter by producing a positively charged ion and a fast-moving electron (Figure 2). The ion and ejected electron can react with molecules to produce other ions and free radicals. (Free radicals are molecular fragments with one or more unpaired electrons. They are usually short-lived and highly reactive.)

$$H_2O \xrightarrow{h\nu} H_2O^+ + e^- \rightarrow H_2O^+ + e^-_{aq}$$

$$H_2O^+ + H_2O \rightarrow H_3O^+ + \cdot OH$$

The hydrated electron can react subsequently with water or with a hydrogen ion to form the hydrogen radical and with oxygen to produce the peroxy radical:

$$e^-_{aq} + H_2O \rightarrow H\cdot + OH^-$$

$$e^-_{aq} + H^+ \rightarrow H\cdot$$

$$e^-_{aq} + O_2 \rightarrow \cdot O_2^-$$

Production of hydroxyl radicals ($\cdot OH$), hydrogen radicals ($\cdot H$), and solvated electrons (e^-_{aq}) occurs within 10^{12} seconds.

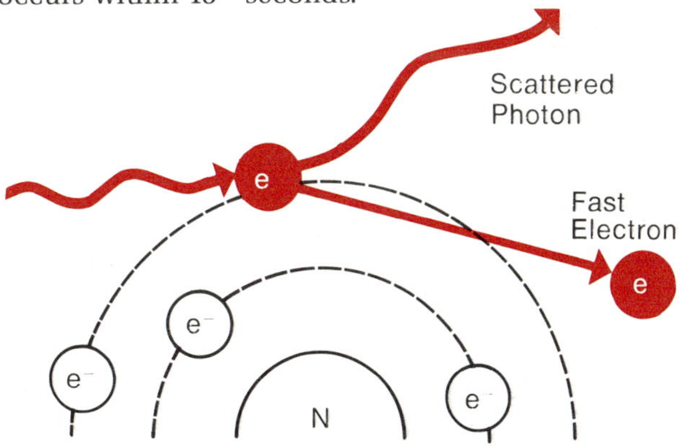

Figure 2. When radiation energy is absorbed, the incident photon interacts with a loosely bound planetary electron of an atom. This photon transfers some of its energy to the electron, which is ejected as a fast electron. The photon then proceeds in a new direction, but with reduced energy.

Biochemical. In tissues, these free radicals interact with organic compounds such as DNA, RNA, enzymes, and other proteins to produce organic radicals:

$$RH + \cdot OH \rightarrow R\cdot + H_2O$$

$$RH + H\cdot \rightarrow R\cdot + H_2$$

In nucleic acids, damage is registered by aqueous radicals interacting with the pyrimidine and purine bases to form additional radicals or by interacting with the phosphate bonds of the sugar phosphate backbone to form either single-strand or double-strand breaks. The direct ionization of organic molecules occurs more frequently with densely ionizing radiation, protons and neutrons, than with X-rays. The actual reaction is similar to that of water:

$$\text{photon} + \text{RH} \rightarrow \text{RH}^+ + e^-$$

$$\text{RH}^+ \rightarrow \text{R}\cdot + \text{H}^+$$

Organic radicals react in two ways. They can react with hydrogen or hydroxyl groups to form more stable compounds, or they can react with oxygen ($RO_2\cdot$). Organic peroxides are very reactive radicals that prevent the restoration of the initial structure of the organic molecule. In the absence of oxygen, however, peroxides are not formed, and many of the ionized organic molecules may return to their initial structure. Unfortunately, some cancer cells have a lower concentration of oxygen than do normal cells. As a result, these cells are relatively resistant to radiation injury. The dose required to kill oxygen-poor cells may be as much as three times that needed to destroy normal cells (Figure 3).

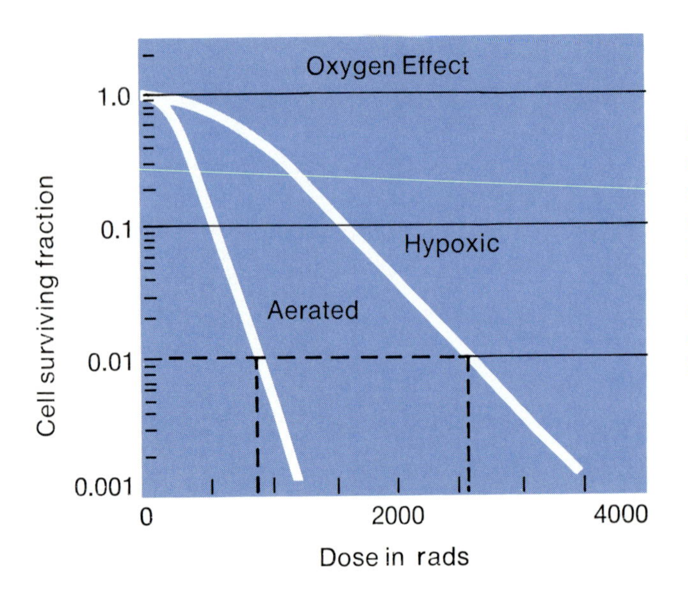

Figure 3. The survival curve for cultured mammalian cells is steeper when irradiation is carried out under aerated conditions rather than under hypoxic or oxygen-poor conditions. This result indicates that the dosage required to kill cancer cells may be as much as three times that required to destroy normal cells.

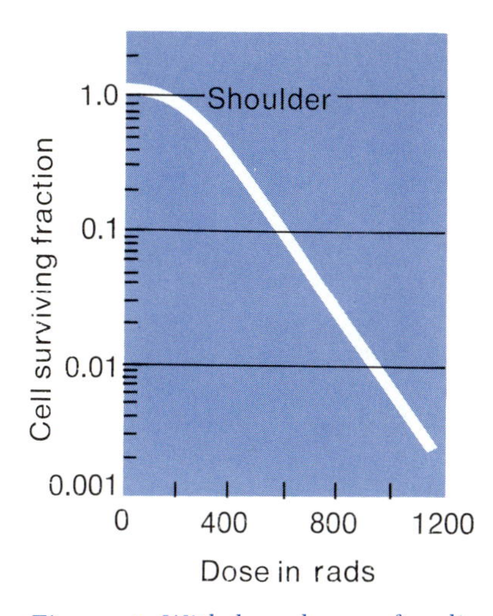

Figure 4. With low doses of radiation, most cells experience damage that is not lethal, but can accumulate to a lethal point. Accumulated sublethal damage is expressed by the shoulder or shallow portion of this cell survival curve. The exponential, or straight, portion of the curve reflects the radiosensitivity of the cells. The steeper the slope, the more sensitive the cells are to radiation. Radiosensitivity is the same among different types of cells, with few exceptions.

Cellular. A minimum number of critical targets in each cell must be hit before the DNA and other targets are inactivated and the cell is killed. Damage to the nucleus causes the cell to lose its ability to proliferate indefinitely. Although it has never been demonstrated conclusively, there is very strong evidence that the DNA molecules are the primary site of permanent radiation damage.

With low doses of radiation, most cells experience damage that is not lethal. However, this sublethal damage can accumulate to a point where it becomes lethal. When the maximum amount of sublethal damage has accumulated, additional radiation will produce a constant fraction of cell death (Figure 4).

Sublethal damage in cells can be repaired if a minimum of four hours is allowed to elapse before delivering additional radiation. If the normal cells can accumulate and repair more sublethal damage than tumor cells, normal cells will be spared when relatively small daily doses of radiation are delivered. For example, because the cells lining the aerodigestive passages can accumulate a larger amount of sublethal damage than cancer cells, multiple small doses delivered four hours or more apart can minimize the destruction of normal aerodigestive lining cells. Patients benefit by reduced soreness of these treated aerodigestive passages and fewer unpleasant effects.

Tissue. The cells of other normal tissues, unfortunately, have a smaller capacity to accumulate and repair sublethal X-ray damage. If this were the only factor involved in determining radiosensitivity, it would be difficult to cure any tumors with radiotherapy without doing equal damage to normal tissue. This is particularly true for tumors containing oxygen-poor cells because normal tissues are generally well oxygenated. Other factors influence the outcome of multiple-dose radiation therapy. Among these are repopulation, reoxygenation, and redistribution of cells within phases of the cell cycle (the position of the cell in the replication cycle affects the cells' radiosensitivity).

Repopulation occurs more readily in normal tissues than in tumors. After treatment, normal untreated cells migrate from the surrounding tissue into the irradiated area. Moreover, normal tissues have better mechanisms to maintain internal stability than do most cancer cells. These mechanisms enable them to increase their rate of proliferation in response to cell death more efficiently than tumor cells.

Reoxygenation of oxygen-poor tumor cells occurs when multiple doses of 150–300 rad are delivered in several weeks. With each radiation dose, tumor cells are killed and the tumor size decreases. Cancer cells that were previously too distant from capillaries for adequate oxygenation are now closer to the capillaries that feed them. As the cells become better oxygenated, they become more radiosensitive.

Minimizing the Oxygen Effect

The destruction of oxygen-poor tumor cells is one of the major challenges of radiotherapy. One way to minimize the oxygen effect is to use densely ionizing radiations as neutrons, protons, and accelerated heavy ions. With these particles, ionizations occur at such close distances that most hydrogen and hydroxyl radicals tend to combine to form H_2O instead of interacting with organic bases to form organic radicals. As a result, the absence of oxygen has less influence on the amount of damage. Instead, damage occurs primarily by direct ionization.

Minimizing the oxygen effect can also be achieved by using molecules that protect oxygenated cells. Radiation protectors spare normal tissues full radiation injury, allowing them to withstand higher doses of irradiation. Many of these protectors are sulfhydryl compounds that compete with oxygen by reacting with free radicals, and thus decrease the production of peroxides. The SH group donates an H atom to the radical site, and thereby restores the DNA molecule to a stable form before the organic radical can react with oxygen to form a peroxide.

Radiation sensitizers are chemical agents that can enhance the lethal properties of ionizing radiation. These agents include electron-affinic compounds that increase the production of free radicals in a manner similar to that of oxygen. They sensitize only oxygen-poor cells and not the well-oxygenated cells. Certain chemotherapeutic agents, halogenated pyrimidines and platinum derivatives, have become useful radiosensitizers. However, their precise method of action is not known.

Radiation Therapy

Radiation therapy is used for two major purposes—to cure cancer and to alleviate some cancer symptoms. Although certain malignancies can be treated effectively with either radiotherapy or surgery, radiation therapy is the treatment of choice for many cancers (*see* list on page 234). In many situations, the advantage of radiotherapy is the ability to preserve the function of an organ, as well as to minimize the mutilating effect of surgery.

Because the effects of radiotherapy are restricted to the volume of irradiation, it is referred to as a loco-regional treatment modality. Cancer patients who receive potentially curative irradiation must satisfy these criteria: (1) cancers localized to an organ with a low propensity for distant spread, (2) an accurately delineated region of involvement, and (3) a cancer volume that may be fully encompassed and treated by a radiation field without an unacceptable risk of permanent damage to adjacent structures. The volume of a cancer is important because the dose needed to eradicate the disease increases as the size of the cancer increases. Volume and dose are reciprocally related for normal tissue. As the irradiated volume of a normal tissue increases, tolerance, the absorbable dose before irreparable damage occurs, decreases. Cell type does not affect radiosensitivity, as discussed earlier, but does influence radiocurability and the total dose necessary to attain long-term tumor control.

Patients who are evaluated for radiation therapy undergo first a biopsy to establish the diagnosis. Then studies, a complete history and physical examination, radiological and isotopic imaging procedures, etc., must precisely determine the anatomic extent and volume of the cancer. When the doses necessary to treat a large cancer would endanger the integrity and function of the surrounding normal tissues and organs, the effect of irradiation may be enhanced by combining it with surgery, chemotherapy, or more recently with heat. By combining treatment modalities, deleterious consequences of either treatment can be minimized because the intensity of each is reduced. These combinations will be discussed after a brief description of some treatment principles in radiotherapy.

Obviously, the defined tumor volume must be completely covered. In addition, the designated dose should be about the same throughout that volume. Normal tissue and organs should receive the least irradiation to the smallest volume possible, with special attention to protecting particularly sensitive organs such as the kidney or lung. Radiation beam directions can vary around a 360° arc, and positions are chosen that direct the irradiation to the cancer while limiting radiation to interposed critical organs. As most malignancies are located internally, megavoltage machines are used. These are more penetrating and cause less skin reaction than low-energy machines. The total dose is usually divided into single daily fractions, given five days a week. This exploits the cancer's greater sensitivity to radiation destruction as the normal tissues repair more of the radiation injury.

Radiation Therapy Hyperfractionation. Rapidly proliferating tumors may occasionally escape the full destructive effect of conventional radiotherapy given once daily. These tumors may repopulate the killed proportion of cells before the next dose of irradiation and remain unchanged or even grow during radiotherapy. Normal tissues have a greater capacity than tumors to repair radiation injury and can do so relatively quickly. Hyperfractionation, the administration of several daily treatments separated by approximately four hours, was studied to take advantage of their resiliance. This treatment technique has proven very successful in some head and neck cancers, and its value is being investigated in several other cancers.

Radiation Therapy Combined with Surgery. Radiation therapy and surgery are combined when the exclusive use of either method would cause unacceptable impairment of organ function or a high rate of failure to control localized cancer (*see* list on page 234). Failure to control local disease with surgery usually results either from undetectable spread of tumor cells to surrounding tissues or to the lymphatic system or from tumor cells seeding the wound, that is, falling from the tumor into the remaining tissue at surgery and subsequently multiplying. With some tumors, unacceptably high doses of irradiation would be required to cure clinically evident disease. Such tumors can be successfully treated by surgical removal of the bulk of the tumor followed by irradiation of residual microscopic disease.

Optimal sequencing of radiotherapy and surgery has not been defined. Preoperative or postoperative radiotherapy both have advantages and disadvantages. Preoperative irradiation decreases the number of viable tumor cells and thereby decreases the chances of tumor cells seeding the wound. The tissues for irradiation are well oxygenated, and therefore conditions for reoxygenation are better than for postoperative radiotherapy. Considerations of tissue alterations with respect to fragility, blood supply, and wound healing strongly influence the total radiation dose and fields used. On the other hand, postoperative radiotherapy allows larger radiation doses and the determination of the exact stage of the cancer. With exact staging, patients with surgically curable disease may avoid unnecessary radiation therapy. However, scar tissue after surgery can decrease the blood supply to some tumor cells; then they are fairly oxygen-poor and thus more radioresistant.

Another newer approach to circumventing the seeming contradiction between maximizing the tumor dose while minimizing normal tissue irradiation is achieved by intraoperative radiotherapy. Here the tumor or tumor bed is isolated and normal tissue is removed from the treatment site as much as possible before irradiation. In this way, a relatively large cumulative dose can be administered to the tumor while a significantly smaller dose is received by surrounding normal tissue.

Radiation Therapy and Chemotherapy. Certain cancers recur locally when treated with chemotherapy; other cancers develop distant metastases when treated locally with radiotherapy. These tumors are best treated with a combination of radiotherapy and chemotherapy (see list). In certain cancers, the predisposition for developing distant metastases as well as local recurrence requires a combination of surgery, radiotherapy, and chemotherapy to control the disease. When surgery is used to remove the primary tumor and radiotherapy and chemotherapy are combined to control disseminated disease, survival rates are greatly improved. In addition, the combination of treatments allows for better preservation of organ function with less extensive surgery and lower doses of radiotherapy. Certain chemotherapy (halogenated pyrimidines and platinum derivatives) can act as an enhancing agent to improve the efficacy of radiotherapy. This radiosensitizer role of the chemotherapy allows some cancers to be treated adequately without extensive surgery.

Radiation Therapy and Hyperthermia. Even though both radiotherapy and hyperthermia can induce tumor responses when administered alone, results have been superior with combined hyperthermia and radiotherapy. Effective techniques for localized hyperthermia generated by external electromagnetic or ultrasonic applicators have been developed only for relatively superficial tumors. The goal is to achieve fairly homogeneous hyperthermia to temperatures of 43–45 °C, and it appears

Treatment Approaches for Various Types of Cancer

Radiation

Hodgkin's disease
Early non-Hodgkin's lymphomas
Cancer of the nasopharynx
Testicular cancer (seminoma)
Cervical cancer
Medulloblastoma
Pinealoma
Optic nerve glioma
Prostatic cancer
Skin cancer
Localized head and neck cancers

Radiation/Chemotherapy

Ewing's sarcoma
Small cell cancer of the lung
Advanced Hodgkin's disease
Acute leukemia in children
Non-Hodgkin's lymphomas
Cloacogenic and squamous anal cancer
Esophageal cancer
Advanced head and neck cancer

Radiation/Surgery

Head and neck cancer
Early breast cancer
Testicular cancer
Rectal cancer
Bladder cancer
Endometrial cancer

Radiation/Surgery/Chemotherapy

Advanced ovarian cancer
Neuroblastoma
Rhabdomyosarcoma
Advanced breast cancer
Wilm's tumor

that minimum tumor temperatures are most strongly correlated with tumor response.

Successful results of using combined radiotherapy and hyperthermia have been reported in malignant melanoma, squamous cell cancers arising from the head and neck region, and chest wall involvement by breast cancers. However, the full potential of combined radiotherapy and hyperthermia has not been realized.

Current research efforts are directed toward improved hyperthermia of deep-seated tumors, combined with sophisticated radiotherapy to achieve a well-localized region of high-dose radiation. Whole-body hyperthermia is another area of current research that will be combined with radiotherapy or chemotherapy. There is interest in developing modulating drugs for hyperthermia to increase heat absorption or cellular responses. Studies are already underway to define the optimal sequencing of radiotherapy and hyperthermia.

Suggested Reading

Hall, E. J. *Radiobiology for the Radiotherapist;* Second Edition; Harper and Row: New York, 1978.

Little, J. B. "Cellular Aspects of Ionizing Radiation." *New England J. Med.* **1968,** *278,* 308–315.

Moss, W. T.; Brand, W. N.; Bottifora, J. *Radiation Oncology: Rationale, Technique, Results;* Fifth Edition; C. V. Mosby: St. Louis, 1979.

Proc. Interaction of Radiation Therapy and Chemotherapy , September 28–October 1, **1986** .

Perez, C. A.; Taylor, W. J. "Symposium on Hyperthermia." *Cancer Res.* **1984,** *44,* 4706s–4908s.

CHAPTER 14 Immunotherapy

Ariel C. Hollinshead

Immunotherapy is a term used to define many different types of cancer therapy that rely on manipulation of the patient's immune system. All forms of immunotherapy are based upon two premises: (1) that there is some biological, chemical, or physical difference between cancer cells and normal cells; and (2) that this difference may be recognized by the host's immune system as a signal to increase the activity of immune system components against cancer and thereby kill the cancer cells. Recent advances in immunology have identified particular cells and cellular products as being of special interest. The idea of immunotherapy is based upon manipulation of the differences between normal and cancer cells and of the changes that occur in the immune system of the cancer patient.

The Immune System

It is fair to ask why the immune system does not get rid of cancer cells at the beginning. The answer appears to be that a normal healthy immune system often does destroy cancer cells in the body. Many normal individuals are able to kill any abnormal cells that develop. Other people get tumors and experience what is called spontaneous

1420–4/88/0237$06.00/0

cure. In other words, their immune system allows a tumor to grow, and then something happens that permits the immune system to recognize the presence of these cells. The immune system mounts an attack upon the tumor and destroys it. We really don't understand fully just how spontaneous remissions occur. We do know that spontaneous remissions occur more frequently with certain types of cancer than they do with other types.

Many exciting studies are underway to define the changes in the immune system that permit either the initiation or the development of cancers. Other studies are attempting to define how cancer cells grow, increase in number, and spread in the body. This chapter will not cover all of these studies, but rather will be confined to actual therapy studies, examples of the major forms of immunotherapy that have been applied to the treatment of patients with major forms of cancer.

Response of the human immune system has been discussed by Fahey and Liu in Chapter 9 of this book. In brief, antigens, substances foreign to the body, enter the body and are trapped and engulfed by the macrophages. A macrophage internalizes an antigen and displays the molecule on its surface; a chemical message released by the macrophage is recognized by circulating T-lymphocytes responsive to this specific antigen. The T-lymphocytes specific to this antigen proliferate; they send signals to activate specific B-lymphocytes and to summon additional macrophages.

The T-lymphocytes and B-lymphocytes increase in numbers and produce both memory cells and effector cells. These memory lymphocytes live a long time in the body and continue to circulate; they are ready to mount a defense if the specific antigen is reintroduced later. Effector lymphocytes are responsible for the immediate response and destruction of these antigens. The B-lymphocytes increase in number and produce plasma cells that secrete into the bloodstream a specific antibody to the antigen. These antibodies bind to the surfaces of antigens, and thus allow the macrophages to recognize and consume the antigens. This response is known as a humoral (bloodstream) response.

As the T- and B-lymphocytes proliferate, they also recruit nonspecific destructor cells, which are triggered by the humoral response and may move into the area to engulf and destroy antigens. In the cell-mediated (tissues) response, not dependent on antibodies, noneffector T-cells are activated and directed to the location of the antigen by specific factors called lymphokines. Lymphokines, soluble products of lymphocytes, are also called mediators of cellular immune reactions. Lymphokines are released into the area by sensitized T-lymphocytes. Another group of cells, natural killer (NK) cells, is difficult to define; they share certain cell-surface antigens with both B- and T-cells. NK-cells are large granular lymphocytes defined operationally for cytotoxicity to certain so-called NK-susceptible tumor cell lines.

Tumor-Associated Antigens

Immunotherapy is based on the idea that the cell surfaces of cancer cells possess differences that the body can recognize and exploit. These are called *tumor-associated antigens* (TAA). TAAs are cell-membrane proteins that are recognized in patients having similar tumors and are not distinct to the patient in whom the tumor arose. Variances of TAA have been found (heterogeneity), so the original tumor may have different TAA from cells seen in metastasis. Such a situation would explain why some metastases (the spreading of the tumor to other organs in the body) resist killing by the immune system: the memory and effector cells had not been educated to recognize all possible TAA variations.

The surfaces of cells are the key to all thoughts about cancer-cell killing. The role of the cell surface in immune recognition is important. The conclusion that TAAs exist can be made with certainty. Resistant lines of cancer cells may exist that cannot be killed with our present immunotherapy. This problem implies the need for additional forms of immunotherapy, as well as for additional work on immunotherapy using TAAs. A partial list of other approaches and components still under development is shown below.

Some Components and Applications Under Study for Possible Use in Immunotherapy

1. Modified growth factors and cell growth factor receptor antibodies
2. Tumor necrosis factor, a protein secreted by endotoxin- or fibrin-activated macrophages
3. Thymosin, a protein made in the thymus that causes maturation of T-cells
4. Monoclonal antibodies to tumor antigens and to tumor antigen genetic components
5. Radiolabeled or drug– or toxin–monoclonal antibody immunoconjugates
6. Intralymphatic immunotherapy
7. Cytokines and monoclonal antibody combination therapy
8. Activated natural killer cells containing a degraded form of an adhesion molecule called laminan
9. Activated macrophages
10. Liposome-encapsulated agents
11. Antisense DNA. Single-stranded DNA complementary to a relevant oncogene for prevention of transcription
12. Interferons, transfer factors, epidermal cell-derived activating factor, lymphotoxin, lymphoregulin, lymphoblastoid cell-produced macrophage activating factor, Tuftsin, tumor angiogenesis factor, and a whole host of other candidate materials for immunotherapy applications
13. Anti-idiotype and chimeric antibody vaccines

Specific Active Immunotherapy

The type of immunotherapy that utilizes tumor-associated antigens is designated *specific active immunotherapy*. The major steps are shown in Figure 1. Many scientists in the past suggested the biological existence of tumor-associated antigens. However, it remained a theory until scientists who used biochemical techniques could identify, isolate, and separate TAA. Pure TAA was tested first in tumor cell cultures; then in virus-induced and carcinogen-induced animal tumors; and finally in patients with colon cancer, lung cancer, and other forms of human solid tumors. These studies are reviewed elsewhere (Hollinshead et al., 1979).

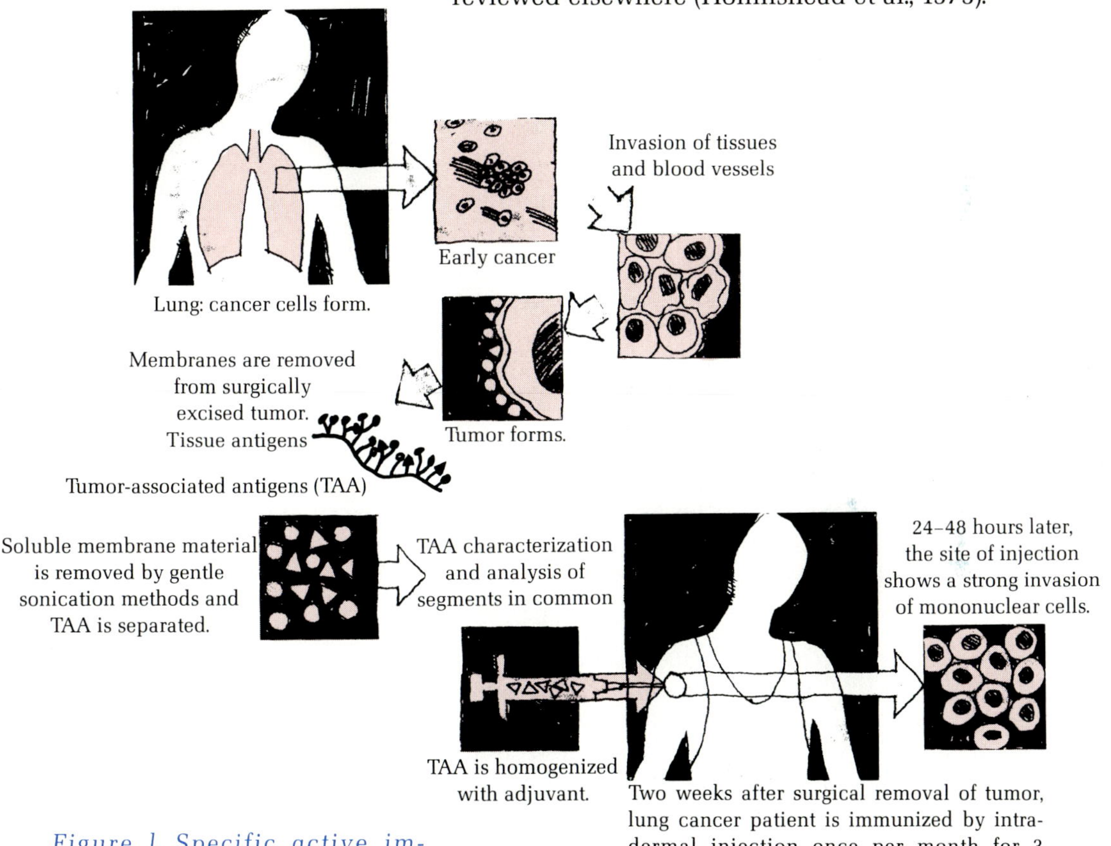

Figure 1. Specific active immunotherapy.

Individual TAAs, which make up 2–3% of the soluble cell membrane proteins, are often masked on the whole cell. Not all the cells in the heterogeneous tumor cell population contain these antigens, and there are quantitative differences in amounts of TAA obtainable from tumor to tumor. Thus, it makes sense to purify and to concentrate known amounts of TAA. Stronger concentrations of pure

TAA, free from interfering substances, induce strong cell-mediated immunity (CMI). Primary tumors of the same histologic type have been studied in large numbers. Only those TAAs that exist in at least 75% of the primary (initial) tumors for a given cancer type and produce an effective CMI response for a given quantity of protein are selected for further studies. TAAs should be free from possible harmful cell products, such as major tissue antigens found on normal cells, nucleic acids, and viral or bacterial products. In certain types of cancers, more than one TAA is utilized in order to produce the strongest cell-mediated immune response for a given concentration of antigen. Thus, an exhaustive amount of cross-testing is performed in patients and in the laboratory, by using different antigens singly and in combination with other selected antigens, in order to develop the strongest combination (called synergism) (Hollinshead et al., 1988). The aim is to destroy the maximum number of cells showing TAA surface variations.

Specific active immunotherapy is most effective when the primary tumor has been removed or reduced in size by surgery. Lung cancer provides an example of the way in which specific active TAA immunotherapy is applied. Doctors at Roswell Park Memorial Institute (Hollinshead et al., 1988) have conducted studies with purified lung TAAs in patients with early-stage (I or II) lung cancer following surgery. As shown in Figure 2, these patients received two forms of immunotherapy. One group of patients received purified TAA well homogenized with an *adjuvant* (a material that nurtures an immune response and permits very slow release of antigen) given intradermally once per month for three months, after which the patients received no further immunotherapy. In another group of this trial, the patients received adjuvant only in one site once per month for three months, with smaller doses of TAA given frequently at another site. This latter treatment is called *separate-site immunotherapy*. A third group of patients, who functioned as controls, received surgery but no further treatment. The patients in these groups were initially randomized so that no

Isolation, Purification, Identification, and Character of TAAs

It was necessary to devise methods that separated cell membranes from the rest of cell, and to develop gentle methods for release of the soluble cell membrane proteins that would not harm their immunogenicity (immune activity in the body) or their antigenicity (activity as antigens, intact and capable of inducing immune responses). A review of the methods and steps involved in the isolation, purification, identification, and characterization of these antigens, as well as some of the humoral and cell-mediated immunity mechanisms involved in specific active immunotherapy, is presented elsewhere (Hollinshead et al., 1988). In brief, the separated cancer cell membranes are subjected to low-frequency sound waves (sonication). The released soluble proteins are then further separated by using selected methods that ensure retention of biological activity, such as ultrafiltration on chromatographic columns (Sephadex), separation by charge using gel electrophoresis techniques, and in some instances separation according to charge by ion-exchange chromatography or according to affinity for monospecific or monoclonal antibodies by affinity chromatography.

Separated antigens are then prepared in ultrapure form by using techniques known as *isotachophoresis* (the principle is based on migration in an electric field of ion species of the same sign, all having a common counter ion) or by further affinity chromatography, in which purification is carried out by utilizing the affinity of a selected antibody for a selected TAA. Identification and characterization of the antigens is carried on throughout purification and ultrapurification, by comparing electrophoresis profiles with profiles of proteins of known molecular weight and activity, densitometry profiles, behavior by immunodiffusion–immunoelectrophoresis against a battery of hyperimmune and immune and control sera and prepared antibodies, by enzyme immunoassays to characterize TAA with tumor-related and control sera, and by other tests. Cell-mediated immune (CMI) responses are measured in a laboratory by using specific lymphocyte stimulation assays, in which the antigens are tested against lymphocytes from patients with the same form of cancer, as well as lymphocytes from controlled patient populations. The in vivo CMI determinations are performed by using delayed hypersensitivity reaction skin testing in titration assays in the related cancer patient and in matched control cancer patient populations.

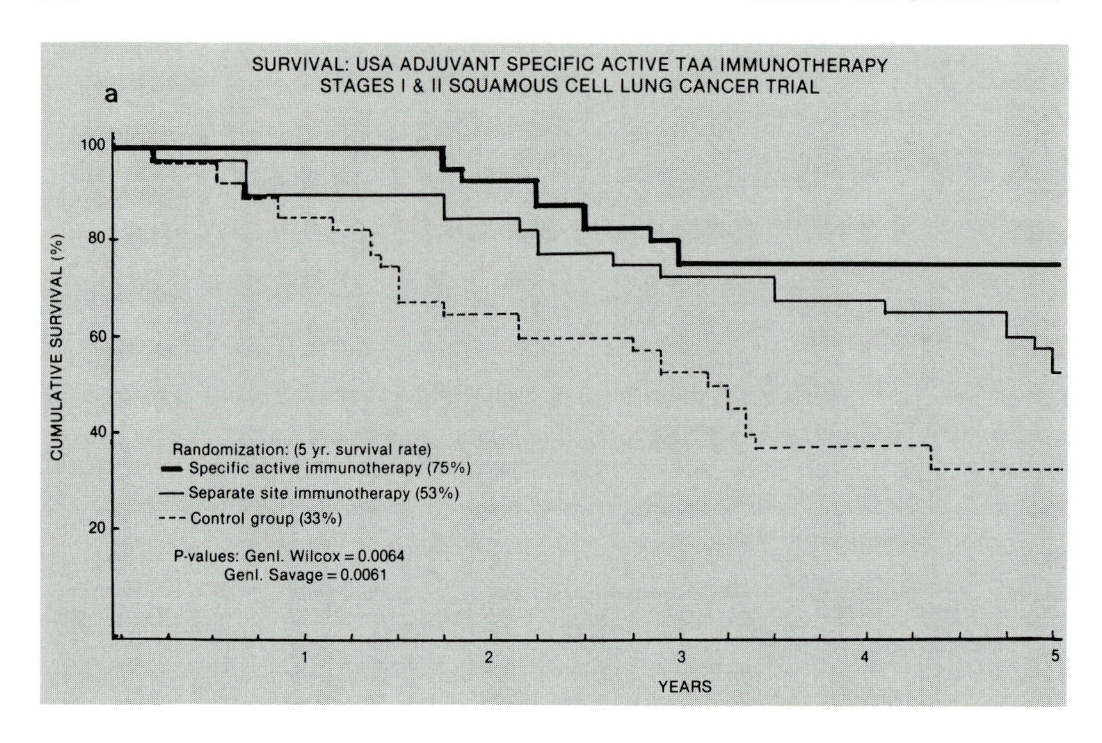

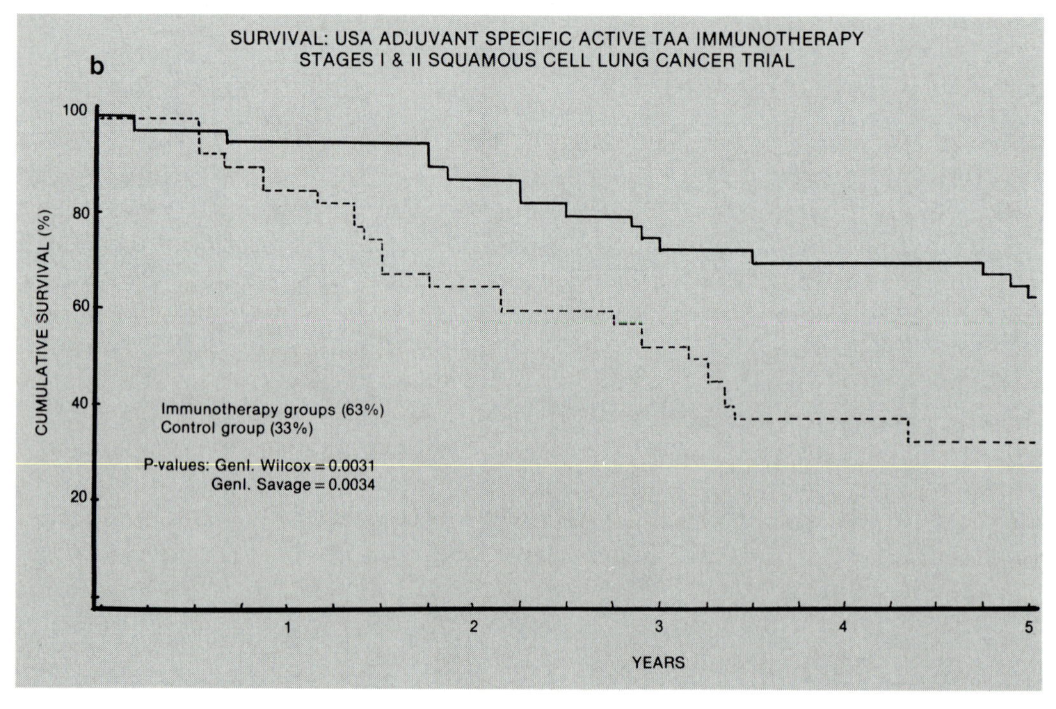

Figure 2. Survival rates with specific active immunotherapy.

bias would be present to any of the three arms of the trial.

As shown in Figure 2a, the five-year survival rate of patients receiving specific active immunotherapy was 75%, with a plateau reached by the third year. Patients on separate-site immunotherapy survived at a rate of 53% for five years, whereas only 33% of the control patients survived five years (Hollinshead et al., 1988). This result means that the single-site therapy was best. However, if both forms of immunotherapy, single-site and separate-site, are combined, as shown in Figure 2b, the survival of patients in the immunotherapy groups is 63%, as compared with 33% in the control group. These early-stage lung cancer patients who received surgery plus immunotherapy doubled their chances of survival, as compared with control groups who received surgery only. This is a major advance in the treatment of solid-tumor cancers, especially because no earlier form of chemotherapy or radiotherapy has substantially affected the five-year survival rate of patients with lung cancers.

Nonspecific Immunotherapy

Can we improve upon the efficacy of immunotherapy? There are other types of immunotherapy besides specific active immunotherapy. In one approach, *nonspecific active immunotherapy*, various agents are studied to determine whether systemic introduction of these agents will stimulate general immune responses in cancer patients. This was one of the earliest approaches to immunotherapy. It was pioneered by William Coley, who first injected bacterial toxins into a 16-year-old boy with inoperable cancer in 1893. He reasoned that, because the immune system of cancer patients is sometimes suppressed, such general stimulation with bacterial toxins might be useful in restoring immune competence.

In 1963, George Mathe administered a live tuberculosis vaccine, BCG, to patients with acute lymphoid leukemia in a similar hope of stimulating the immune system to attack cancer cells. Other types of BCG, as well as other bacterial products, have been studied by other investigators in a large number of clinical trials. The results were not promising. However, in certain individuals, these approaches occasionally worked. Immune responses to the patients' own tumor-associated antigens may occur, although at a lower level than when the patient is stimulated by specific active immunotherapy. In the few instances in which nonspecific immunotherapy appears to work, the patient's own immune system might have been stimulated by chance.

Regional Immunotherapy

Another form of immunotherapy is called *regional immunotherapy*. In this approach, bacterial products are injected into the lesion itself. Although this form of regional immunotherapy might cause easily accessible skin lesions to disappear, the results were disappointing. This form of therapy did not affect either systemic cancers or subcutaneous lesions.

Adoptive Immunotherapy

In *adoptive immunotherapy* the peripheral blood lymphocytes are sensitized outside of the body to various cell products and then returned to the patient so as to transfer the effect of the treated cells to the patient. This is called adoptive transfer.

Most of the products used in adoptive transfer come from cells that are involved in the humoral or cell-mediated immune system. Some of these substances, called intermediary cell products, are lymphokines; they are often designated as biologic response modifiers. Some of these products are used to inhibit cancer cell growth directly by their toxic effect on cancer cells, and some are used to augment natural killer cell activity against the tumor. Because most of these studies are still in their infancy, only the most promising initial clinical studies will be described here.

In one study (Rosenberg et al.), it was possible to increase the body's response against cancer to the extent that certain solid, established, growing tumors were made to shrink. Occasionally, tumors disappeared. This result also occurs with specific active immunotherapy (Hollinshead, 1986). The adoptive immunotherapy trial was based upon the use of a substance called interleukin-2, which is a T-cell growth factor. In a sense, interleukin-2 (IL–2) could be called a hormone. It has been possible to produce interleukin-2 artificially by using recombinant gene technology (Rosenberg et al., 1985).

Scientists raised a host of killer cells in vitro in the presence of IL–2. These killer cells can recognize cancer cells, yet spare normal tissues of the patient. The initial work at the National Cancer Institute involved raising huge numbers of killer lymphocytes in laboratory cultures and infusing such cells into cancer patients. This step was accomplished by removal of human blood lymphocytes. Their incubation in IL–2 resulted in the production of interleukin-2-activated killer cells, called LAK cells (lymphokine-activated killer cells) (Rosenberg et al., 1985). Patients received systemic administration of IL–2 alone and in conjunction with systemic administration of these sensitized expanded lymphoid cells (LAK cells). Of 80 patients with advanced cancer who received IL–2 alone, 4 patients with renal cancer have undergone complete regression. Of 137 patients with advanced cancer who received LAK/IL–2, 12 patients (7 renal, 3 melanoma, 1 colon, and 1 lymphoma patient) have undergone complete remission. No effect was seen in patients with sarcoma, lung cancer, breast cancer, and other forms.

Of the 16 patients showing improvement (out of a total of 217 in the project), 13 remained without disease 2–31 months (Rosenberg, 1988). These were not "educated" killer cells in the sense that they had been exposed to any specific tumor antigens, although some predilection for attacking cancer cells had been noted. The toxicity was very severe. Many groups are studying the efficacy of this form of therapy.

It was quickly hypothesized that if one could

take lymphocytes that are "educated" by TAA to attack cancer cells specifically, increase their numbers into the billions, and infuse them back into the patient from which they arose, then toxicity would be reduced and efficacy enhanced. This method would require surgical removal of the tumor and laboratory culture to produce enormous numbers of the lymphocytes found in the tumor. Such tumor-infiltrating lymphocytes could only be given safely to the patient from which they came.

The combination of specific active immunotherapy and adoptive immunotherapy may be very attractive. One could theorize that under the very potent influence of the hormone interleukin-2, the killer cells that recognize TAA can be induced to grow to numbers needed to kill the cancer target cells. An explosive growth of these killer lymphocytes can be induced in laboratory culture conditions, or perhaps in the patient bearing the tumor. Thus, a combination of these two forms of immunotherapy may be advantageous.

Passive Immunotherapy

So far we have discussed antigens and intermediary cell products, in particular a lymphokine called interleukin-2. But what about antibodies? In many laboratories, workers are cloning cells that have been primed to produce particular antibodies, preferably specific to some characteristic that is unique to the cancer cell surface. These are called *monoclonal antibodies*. In the large clinical studies thus far conducted, monoclonal antibody therapy alone has not been effective (Campbell et al., 1987). Because the body's immune system is not activated by injections of antibodies, monoclonal antibodies must be administered continually for months and years. This form of therapy is called *passive immunotherapy*. A number of studies are underway to hook monoclonal antibodies to drugs, toxins (including some very similar to Coley's toxins), and other anticancer agents with the hope that this delivery system will direct the activity of these agents to the cancer itself and not to other tissues. To date, there are no results from large clinical trials, and we must await the completion of the studies. One interesting experimental approach, shown in Figure 3, is research involving antibodies produced to antibodies. The work involved is complex, but future applications hold much promise (Kennedy et al., 1986).

Summary

Specific active immunotherapy appears to be an effective approach to be added to the armamentarium of therapeutic approaches, especially surgery. It may be useful in combination with those forms of chemotherapy and radiotherapy that are effective for particular types of cancer. These combined

This is an example of the way in which experiments are done to develop laboratory-produced immunogens that mimic tumor antigens. There are two forms of these that may have promise:

1. ANTI-IDIOTYPE ANTIBODY VACCINE—An idiotype is the antigenic determinant present on the variable site ( ) of the antibody and is the surface site, or idiotype area, that can elicit an antibody response.

2. CHIMERIC ANTIBODY VACCINE—In this method, the V region is generated by the mouse and the rest of the antibody is human in nature. Whether or not it is necessary to use this form will depend on experiments to determine whether multiple injections of a product of mouse origin will produce anaphylaxis when it is studied in controlled trials in humans.

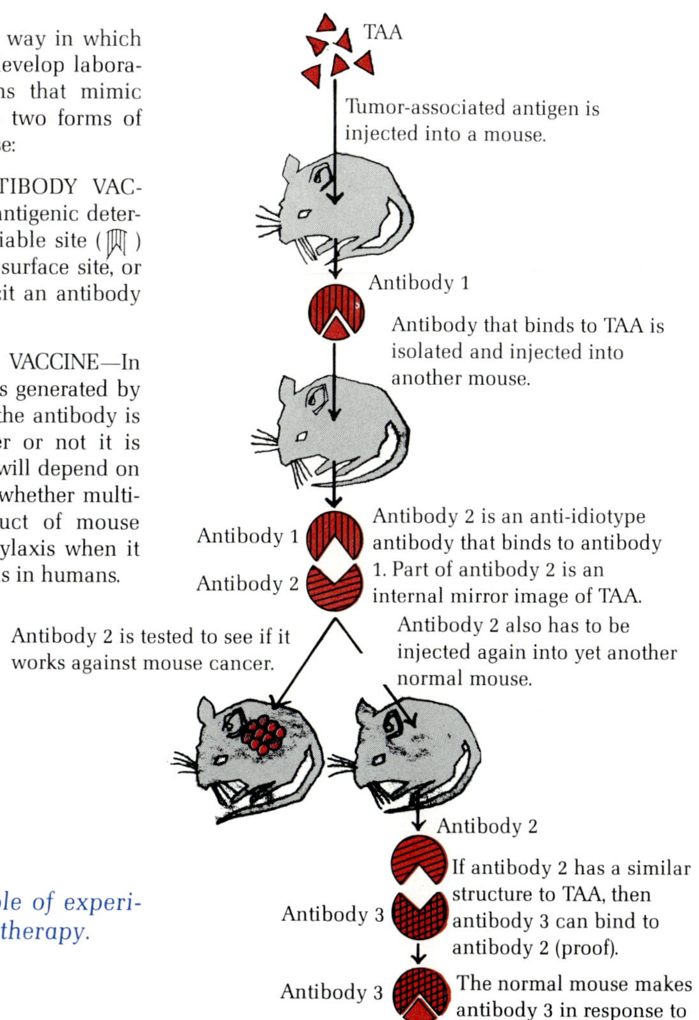

Figure 3. An example of experimental immunotherapy.

therapies must be studied in randomized trials in a large group of patients in order to ascertain whether they are more or less effective than therapies used previously. Immunotherapy has a strong role to play in the future, and there is no question that additional strategies, based upon our evolving understanding of the immune system, will result in even more effective treatments in the years ahead.

References

Campbell, A. M.; Whitford, P.; Leake, R. E. "Human Monoclonal Antibodies and Monoclonal Antibody Multispecificity." *Br. J. Cancer.* **1987,** 56, 709–713.

Hollinshead, A.; Stewart, T.; Takita, H. "Tumor-Associated Antigens; Their Usefulness as Biological Drugs." In *Lung Cancer: Progress in Therapeutic Research;* Muggia and Rozencweig, Eds.; Raven Press: New York, 1979; pp 501–520.

Hollinshead, A.; Takita, H.; Phillips, T.; Drell, D; Fudenberg, H.; Stewart, T.; Raman, S. "Humorocellular Immune Correlates of Clinical Responses of Lung Cancer Patients Receiving Adjuvant Specific Active Immunotherapy Versus Immunochemotherapy." *Cancer,* in press (1988).

Hollinshead, A. "Immunotherapy Trials: Current Status and Future Directions with Special Emphasis on Biologic Drugs." In *New Perspectives in Immunotherapy,* Hadden, J.; Spreafico, F., Eds. Springer Seminars in Immunopathology, Miescher; Muller–Eberhard, Eds.; Springer–Verlag: Berlin, 1986; pp 85–104.

Kennedy, R. C.; Melnick, J. L.; Dreesman, G. R. "Anti-Idiotypes and Immunity." *Scientific American* **1986,** *255* 48–56.

Rosenberg, S.; Lotze, M.; Muul, L.; et al. "Observations on the Systemic Administration of Autologous Lymphokine-Activated Killer Cells and Recombinant Interleukin-2 to Patients with Metastatic Cancer." *N. Engl. J. Med.* **1985,** *313,* 1485–1495.

Rosenberg, S. "Immunotherapy of Cancer Using Interleukin 2: Current Status and Future Prospects." *Immunology Today* **1988,** 9, 58–62.

Glossary

Glossary

Adduct: product of a chemical reaction

Adhesion plaque: areas of cell that come in close contact with a surface; stress fibers terminate at these specialized regions of the cell membrane; may allow communication between cell surface and nucleus

Alleles: pair of genes at the same position on both members of a pair of chromosomes and conveying characters that are inherited alternatively

Angiogenesis: the formation and differentiation of blood vessels and capillaries

Antibody: protein produced by B-lymphocytes (B-cells) in response to antigen; reacts specifically with its complementary antigen

Antigen: a protein, toxin, or other substance, usually of high molecular weight, to which the body reacts by producing antibodies

Antiserum: immune serum containing antibodies active chiefly in destroying a specific infecting virus or bacterium or reacting with an experimentally introduced antigen

Athymic: animals deprived of their thymus by surgery or genetic defect; *see* T-cells

B-cells: antibody-producing cells

Basement membrane (lamina): thin layer of extracellular matrix forming a boundary between epithelial cells and underlying connective tissue

Carcinogen: any substance that produces cancer

Carcinoma: cancer originating from epithelial cells

Chemotherapy: use of chemicals to kill cancer cells with minimal damage to normal cells

Choriocarcinoma: malignant tumor derived from

chorionic tissue arising spontaneously in the testis, in the ovary following pregnancy, or extragenitally in the mediastinum

Chromosome: threadlike structures in plant and animal nucleii consisting of DNA complexed with an equal mass of proteins; carry linearly arranged genetic units; visible by light microscope during cell division

Clone: descendant derived asexually from a single parent

Cocarcinogen: substance that is harmless by itself, but enhances the effect of a carcinogen when applied with it

Collagen: fibrous protein found especially in connective tissue

Complete remission: relatively prolonged disappearance of symptoms

Cytogenetics: study of the behavior of chromosomes and genes in cells with regard to heredity and variation by means of the light and electron microscope

Cytoplasm: protoplasm of a cell, outside the nucleus and surrounded by the cell membrane

Cytoskeleton: oriented framework of complex protein fibers, believed to be responsible for the mechanical properties of protoplasm

Cytotoxic: poisonous to cells

DNA: deoxyribonucleic acid; carrier of genetic information; a linear polymer present in chromosomes of cells and in some viruses

Degranulation: release of cellular granules

Densely ionizing radiation: large-particle radiation that deposits a high dose of energy directly in the tumor, with a low dose to the surrounding normal tissue

Disjunction: separation of members of chromosome pairs as each member moves to opposite poles during cell division

Elastin: protein; principal component of elastic fibers

Electrophile: compound that reacts easily with any available electrons

Electrophoresis: movement of colloidal particles suspended in a fluid, caused by application of an electric field; allows separation and purification of proteins

Embolus: a clot or mass of particles foreign to the bloodstream

Endothelium: epithelial layer of cells lining the heart and the vessels of the circulatory system

Enzyme: protein that serves as a catalyst of biochemical reactions

Epigenetic inheritance: processes beyond the sequences dictated in DNA that control gene expression, such as modulation of transcription, protein synthesis, and apportionment of proteins in the cell

Epithelium: cellular tissue covering surfaces, forming glands, and lining most cavities of the body

Extravasation: pouring out or eruption of a body fluid from its proper channel into the surrounding tissue

Fibrin: fibrous insoluble protein that forms the structure of a blood clot

Fibroblast: most common cell found in connective tissue; involved in wound healing

Free radicals: atoms or groups of atoms possessing an odd (unpaired) electron

Genome: genetic endowment of a species

Germinal: refers to characteristics present in germ cells (sperm and egg) as distinct from somatic cells

Glycolipid: complex lipid containing carbohydrate residues

Glycoprotein: compound containing a protein and a carbohydrate group(s)

Growth autonomy: capacity for uncontrolled growth

Growth factor: conditions such as space and nutrients that govern cell growth

Heterozygote: plant or animal with two different alleles at a single position on a chromosome and hence not breeding true to type for the particular character involved

Histocompatibility: compatibility between tissues of a graft or transplant and tissues of the body receiving it

Homeostasis: tendency of an organism to maintain normal internal stability

Humoral immunity: production of antibodies by B-lymphocytes (B-cells)

Hybridoma: hybrid myeloma formed by fusing myeloma cells with lymphocytes that produce a specific antibody; individual cells can be cloned to produce large amounts of identical antibody

Hyperthermia: heat treatment of tumors

Immortal cells: cells that have no limit on the number of times they can divide; show lessened dependence on growth factors to stimulate division

Immunoconjugate: union of an immunogen (a substance that stimulates some degree of immunity) with one of several substances, such as a monoclonal antibody hooked to a drug

Immunology: branch of medicine dealing with antigens and antibodies, especially immunity to disease

Immunotherapy: therapy based on manipulation of the patient's immune system so that it will attack the cancer cells

Intermediate filaments: irregular cytoplasmic fibers intermediate in diameter between microtubules and actin filaments; provide mechanical support to the cytoplasm

Intradermal: within the skin or between the layers of the skin

Ionization: conversion of neutral atoms or molecules into charged particles

Isoenzymes: distinct forms of an enzyme, having different electrochemical states and representing different polymeric states, but having the same function

Karyotype: entire complement of chromosomes present in a cell; observed by photographing chromosomes through a microscope during mitosis

Keratin: sulfur-containing fibrous proteins that form the chemical basis of epidermal tissues such as horn, hair, wool, nails, feathers

Kinase: enzyme that catalyzes phosphorylation reactions

Lesion: injury or other change tending to result in impairment or loss of function

Leukocyte: white blood cell

Lymphocyte: type of leukocyte concerned with immunity, such as B- and T-lymphocytes

Lymphoma: any tumor, usually malignant, of the lymphoid tissues

Lysis: refers to death of cell as a result of destruction of cell membrane, allowing escape of cell contents

Macrophage: large ameboid cell in the reticuloendothelial system; engulfs foreign material

Meiosis: process of nuclear division in which the number of chromosomes normally is reduced in the daughter cells

Metastasis: spread of malignant cells to another part of the body, usually by way of the bloodstream or lymph system

Microfilaments: stress fibers, especially abundant in muscle cells; involved in shape and contraction of muscle fibers

Microtubules: hollow tubelike filaments; involved in cell movement

Mitosis: nuclear division involving exact duplication, so that each of two daughter nuclei carries a chromosome complement identical to that of the parent nucleus

Monoclonal antibody: highly specific antibody produced by hybridoma cells; binds with a single antigenic determinant

Morphology: structure and form of an organism, tissue, or cell

Motility: spontaneous movement

Murine teratocarcinoma: cancerous teratoma in rats or mice

Mutagen: anything that raises the frequency of genetic change above the spontaneous rate

Mutation: permanent change in DNA

Myeloma: primary tumor of the bone marrow, such as neoplasms arising from B-lymphocytes

Necrosis: death or decay of tissue

Neoplasia: formation of a neoplasm or tumor

Neoplasm: a heritably altered, relatively autonomous growth of tissue; growing tumor

Neoplastic transformation: cell growth becomes uncontrolled, with alterations in cytoskeleton;

communication between cell exterior and nucleus seems lost

Normalization: cancer cells will develop normally when subjected to strong developmental influences (e.g., implanted in embryos or treated with inducer substances)

Nuclear envelope: double membrane separating cytoplasm from nuclear contents, including chromosomal DNA

Nucleic acids: found in cell nucleus and in some cytoplasmic organelles; they include DNA (genetic information) and RNA (involved in protein synthesis)

Nucleophile: compound rich in electrons available for reaction

Nucleotide: structural unit of a nucleic acid

Oncofetal antigens: antigens produced during embryonic development; generally repressed during adult life, but may appear in regenerating or neoplastic tissue

Oncogene: cancer-causing gene

Oncogenesis: process of tumor formation

Organelle: specialized structure within cells, having a specific function

Oxygen effect: cells with a high concentration of oxygen are most sensitive to radiation damage; oxygen-poor tumor cells are hard to destroy by radiation without destroying neighboring normal tissue

Parenchyma: specialized epithelial (covering) part of an organ

Pathogen: disease-producing agent

Pathogenesis: Production or development of a disease

Phenotype: observable characteristics of an organism

Phosphorylation: transfer of phosphate group from high-energy donor such as ATP to receptor molecules in biochemical metabolism of the cell

Photon: quantum of electromagnetic energy having both particle and wave behavior

Polymorphism: condition in which a species has two or more different structural forms, such as Rh and other blood groups in humans

Proteoglycan: polyanionic substance linked to a polypeptide chain backbone

Proto-oncogene: normal gene incorporated into the viral chromosome of a retrovirus, frequently altered by mutation, then becomes an oncogene

Provirus: integrated viral DNA

RSV: Rous sarcoma virus; first virus shown to cause cancer in animals

Radiation therapy: use of controlled irradiation to destroy a tumor while causing only minimal damage to surrounding normal tissues or to alleviate some cancer symptoms

Radiation tolerance: absorbable dose before irreparable damage occurs in a given organ or tissue

Recurrence: return of a malignancy after a period of remission

Remission: relatively prolonged lessening or disappearance of the disease symptoms

Retrovirus: RNA viruses that utilize reverse transcriptase to copy the viral RNA sequence into DNA; the resulting DNA is then integrated into the host cell's chromosome

Ribonucleoprotein: conjugated protein in which ribonucleic acid molecules are closely associated with protein molecules

Sarcoma: malignant tumor arising in connective tissue

Seeding the wound: malignant cells can fall from a tumor into the remaining tissue during surgery and then multiply

Senescence: limit to the number of times a normal cell can divide before terminal differentiation; a normal human cell in culture stops dividing after 40–50 cell doublings

Somatic cells: all the cells in the body except germ cells

Sonication: use of the energy produced by low-frequency sound waves to fragment cells for biochemical extraction

Specific active immunotherapy: immunotherapy that uses tumor-associated antigens

Stem cells: undifferentiated cells that give rise to specialized daughter cells during repeated cell divisions; for example, the mature cells of blood and skin are produced by stem cells

Stress fibers: criss-crossed bundles of musclelike intracellular cables

Stroma: supporting framework of an animal organ, made up of connective tissue

Syngeneic: highly inbred animals or identical twins; can exchange skin grafts without rejection

T-cells: thymus-derived lymphocytes that can destroy foreign cells or antigens either by direct attack or by activating other components of the immune system

Teratoma: tumor containing various kinds of embryonic tissue

Terminal differentiation: permanent departure of a cell from the developmental cycle in order to serve a specialized function

Thrombin: enzyme in shed blood that induces clotting

Thromboplastin: lipid and protein complexes in blood that accelerate production of thrombin

Totipotent cell: cell that has the ability to differentiate into a whole organism

Transfection: incorporation of foreign DNA into cultured animal cells by mixing the DNA with the cells

Transformation: process by which a normal cell attains characteristics of a cancer cell

Translocation: fusion of one portion of a chromosome with another chromosome

Tumor-associated antigens: cell-membrane proteins that can be recognized in patients having similar tumors

Tumor angiogenesis factor: substance synthesized by tumor cells that stimulates development of a network of blood vessels to sustain the tumor

Tumor suppression: some cancer cells can be made to develop normally without genetic change; in a different environment (e.g., in culture) they may exhibit cancerous properties

Vascular: of the system of channels for transporting blood or lymph

Vascularization: growth of capillary blood vessels into a tumor

Viscera: organs within the cavities of the body

Xenogeneic: referring to cells, tissues, or organs used in transplantation to a different person or species

Indexes

Indexes

Author Index

Affiliation Index

Subject Index

A

B

C

N

U

V

W

X

Y

Editing and indexing by Colleen P. Stamm
Production by Barbara J. Libengood
Jacket design by Alan Kahan
Managing Editor—Janet S. Dodd

Typeset by Hot Type Ltd., Washington, DC
Printing and paperback binding by The Sheridan Press, Hanover, PA
Clothbound binding by Maple Press, York, PA